MW00835397

A Course in Complex Analysis in One Variable

A Course in Complex Analysis in One Variable

Martin A. Moskowitz

Professor of Mathematics
City University of New York Graduate Center

Published by

World Scientific Publishing Co. Pte. Ltd.

P O Box 128, Farrer Road, Singapore 912805

USA office: Suite 1B, 1060 Main Street, River Edge, NJ 07661

UK office: 57 Shelton Street, Covent Garden, London WC2H 9HE

British Library Cataloguing-in-Publication Data
A catalogue record for this book is available from the British Library.

A COURSE IN COMPLEX ANALYSIS IN ONE VARIABLE

ISBN 981-02-4780-X

Printed in Singapore by Mainland Press

Preface and Acknowledgments

This book was written with two groups of readers in mind. One comprises first year graduate students in mathematics who are either studying for their qualifying exams, or who want to learn the basics of this important subject. Since much of the content of this book originated in a one semester course I gave at the City University of New York Graduate Center, it would be very suitable for such an audience. The second group comprises advanced undergraduate mathematics or science majors. For this purpose there is enough material for a year-long course or, by concentrating on the first three chapters, for a semester course. The exposition is built around the fundamental concept of a holomorphic function in a planar domain and does not address more advanced topics such as Riemann surfaces. Although the material is classical, I believe it has been organized here in an especially efficient manner, getting basic complex analysis into about one hundred and thirty printed pages. Nevertheless, I have striven to relate and contrast the material with that of its sister, real analysis.

My own introduction to the subject occurred many years ago when, as an undergraduate, I read K. Knopp's paperback book [5]; some of the present material is informed by the treatment there. The proof of the Riemann mapping theorem for planar domains derives largely from Conway [1] although in somewhat greater detail than appears in his book. The same is true of holomorphic diffeomorphisms of annuli and

Rudin's book [10]. I have allowed myself the indulgence of including a short final chapter containing some applications of complex analysis to Lie theory and differential topology on which, of course, nothing else depends and which can be omitted if the reader's interests lie elsewhere.

Complex analysis is a beautiful subject, perhaps the single most beautiful and striking in mathematics. It contains completely unforseen results of a dramatic, one might even say magical character. One can imagine the excitement of its founders–first Cauchy, and then Riemann and Weierstrass–as each developed a better and better grasp of the terrain, proving major results one after the other. It is my hope that this compact book will convey to the student some of the excitement and extraordinary character of this subject.

I would like to thank Hossein Abbaspour and Delaram Kahrobaei for making a number of useful typesetting suggestions. In addition, Hossein Abbaspour did the difficult job of creating the diagrams and incorporating them into the text as well, as helping me with the index, and has been tireless in the many duties necessary to bring this book to completion. Rob Landsman was very helpful to the project in dealing with all communications with publishers. I wish to thank Andre Moskowitz for proofreading the manuscript from the point of view of style and Nils Tongring for reading for content. Both made many comments which have considerably improved the exposition. Finally, I wish to express my gratitude to Florian Lengyel for solving a number of the typesetting problems necessary to put this into World Scientific format.

Over the years I have had discussions with several colleagues on matters which have sometimes had a bearing on the subject matter under consideration here. The people who come to mind are Adam Koranyi, Karl-Hermann Neeb, Burton Randol, Richard Sacksteader, Dennis Sullivan, Nils Tongring, and Alphonse Vasquez. If I have left anyone out I apologize. Of course, any errors or misstatements are my responsibility. Finally, I would like to thank my wife, Anita, for her encouragement and forbearance during the course of the project.

Martin Moskowitz, Summer 2001

Contents

Chapter 1

First Concepts

1.1 Fundamentals of the complex field

In this section we give a rapid introduction to the complex numbers themselves. Here we shall assume the reader is largely familiar with this material, or at most needs to have some jogging of the memory.

We denote the complex numbers by $\mathbb{C}$ which is identified with the Euclidean plane $\mathbb{R}^2$ by writing the pair (a, b) as $a + bi$. In $\mathbb{C}$ we add by $(a + bi) + (c + di) = a + c + (b + d)i$ and multiply by $(a + bi) \cdot (c + di) = ac - bd + (ad + bc)i$. Thus $i^2 = -1$. It is easy to check, and this we leave to the reader, that $(\mathbb{C}, +, \cdot)$ is a commutative ring with unit, where $0 + 0i = 0$ is the zero and $1 + 0i = 1$ is the unit. $\mathbb{C}$ is actually a field, that is, every non-zero element has a multiplicative inverse.

To see this let $z = x + iy \in \mathbb{C}$. We define two quantities the norm of a complex number z given by

$$|z| = \sqrt{x^2 + y^2}\,.$$

(here $|z| \geq 0$ and $|z| = 0$ only if $z = 0$) and the conjugate $\bar{z}$ of $z = x + iy \in \mathbb{C}$ by

$$\bar{z} = x - iy\,.$$

A direct calculation shows that $z\bar{z} = |z|^2$. Hence if $z \neq 0$, $z\frac{\bar{z}}{|z|^2} = 1$. Thus $z^{-1} = \frac{\bar{z}}{|z|^2}$ so $\mathbb{C}$ is a commutative field. We denote its multiplicative group by $\mathbb{C}^\times$. Another direct calculation shows $\overline{(z+w)} = \bar{z} + \bar{w}$ and $\overline{(z \cdot w)} = \bar{z} \cdot \bar{w}$. Thus conjugation is an automorphism of the complex field. Using this last remark it follows readily that

$$|zw| = |z||w|,$$

and for $w \neq 0$

$$|\frac{z}{w}| = \frac{|z|}{|w|}.$$

This in turn tells us that if $z \neq 0$ then $\frac{z}{|z|}$ has modulus 1. This process is called *normalizing* z.

The fact that conjugation preserves multiplication also shows that the unit circle

$$\mathbb{T} = \{z : |z| = 1\}$$

forms a subgroup of $\mathbb{C}^\times$ whose elements are characterized by $z^{-1} = \bar{z}$. So in particular $i^{-1} = -i$ and $-i^{-1} = i$. The group $\mathbb{T}$ is compact. It is the group on which Fourier analysis is done. Actually, under the metric defined by

$$d(z, w) = |z - w|$$

the field $\mathbb{C}$ forms a complete metric space. An important ingredient is the so-called triangle inequality

$$|z + w| \leq |z| + |w|.$$

Finally, we mention the polar decomposition of a non-zero complex number. If $z \in \mathbb{T}$, then there is a unique angle θ (actually unique up to an integer multiple of 2π) such that $z = \cos\theta + i\sin\theta$. Of course, this means that more generally, if $z \neq 0$, then $z = |z|(\cos\theta + i\sin\theta)$. This is called the *polar decomposition* of $z \neq 0$. The angle θ is called the argument, or arg of z. (Obviously we must have $z \neq 0$ to have a well defined argument). This decomposition shows $\mathbb{C}^\times$ is a product space, $\mathbb{C}^\times = \mathbb{R}_+{}^\times \times \mathbb{T}$, where the first component of z is its modulus $|z|$ and the second its argument θ. We ask the reader to verify all these facts.

Exercise 1.1 *Let $p(z)$ be a polynomial with real coefficients. Show that the non-real roots occur in conjugate pairs. In particular, if p has odd degree it must have a real root.*

1.2 Holomorphic functions

Since we have a (metric) topology on $\mathbb{C}$ we can talk about open sets. Just as in real analysis, these will be the domains of differentiable functions. Let Ω be a non-empty open set in $\mathbb{C}$. For example, Ω could be $\mathbb{C}$ itself, or $\mathbb{C}$ with say a finite number of points removed (such as $\mathbb{C}^\times$), or the interior of a disk, or a half plane. Analogously to real analysis we consider functions $f : \Omega \to \mathbb{C}$. To distinguish the situation from that of real functions (and as we shall see, it is quite different) we use the term *holomorphic* function.

Definition 1.2.1 Let $f : \Omega \to \mathbb{C}$ and $a \in \Omega$. We say f is holomorphic at a if

$$\lim_{z \to a} \frac{f(z) - f(a)}{z - a}$$

exists. Since if limits exist they are unique, we can give this one the name $f'(a)$, called the derivative of f at a. If f is holomorphic at every point of Ω, we say it is a holomorphic function on Ω. In this case the derivative is denoted by f'. If we wish to take several derivatives we denote by $f^{(k)}$ the kth derivative of f.

Before proceeding we make a few observations. Just as in the case of real functions of a real variable, items 1 and 2 below follow from the definitions and elementary properties of limits in the complex domain. Together they go by the name, the linear approximation theorem. Item 3 follows immediately from item 1.

1. If f is differentiable at a, then

   $$f(z) = f(a) + f'(a)(z - a) + \epsilon(z)(z - a),$$

   where $\epsilon(z)$ tends to zero as z tends to a.

2. If f is defined in a neighborhood of a and

   $$f(z) = f(a) + c(z - a) + \epsilon(z)(z - a),$$

   where c is a constant and $\epsilon(z)$ tends to zero as z tends to a, then f is differentiable at a and $f'(a) = c$.

3. If f is differentiable at a, then f is continuous at a.

Notice that, just as in the real case, being holomorphic is a local property. Moreover, if we have two functions f and g both differentiable at a, then

1. $f \pm g$ is differentiable at a and $(f \pm g)'(a) = f'(a) \pm g'(a)$.
2. Similarly, fg is differentiable at a and $(fg)'(a) = f'(a)g(a) + f(a)g'(a)$.
3. If, in addition, $g(a) \neq 0$, then $\frac{f}{g}$ is also differentiable at a and $(\frac{f}{g})'(a) = \frac{f'(a)g(a) - f(a)g'(a)}{g(a)^2}$.

We ask the student to check all these facts by looking at the analogous material in a calculus book.

Since the derivative of the constant function is clearly zero, from the second of these formulas it follows that the derivative of a constant times f is $c \cdot f'$. Also from the second formula, we see by induction that the function $f(z) = z^n$ is holomorphic on all of $\mathbb{C}$ with $f'(z) = nz^{n-1}$. Hence by the first formula, together with the above, all polynomials are holomorphic on $\mathbb{C}$. Also, if f is a polynomial of degree n, then taking $n+1$ successive derivatives gives zero. Functions that are holomorphic on all of $\mathbb{C}$ have a special name; they are called *entire*.

The third formula gives examples of holomorphic functions on domains other than $\mathbb{C}$. Let $f(z) = \frac{p(z)}{q(z)}$ where p and q are polynomials. These are the so-called rational functions. By the third formula these are holomorphic everywhere on $\mathbb{C}$, except at the zeros of q.

Another fact whose statement and proof are completely analogous to that of the real case is the chain rule. If $f : \Omega \to \Omega' \subseteq \mathbb{C}$ is holomorphic at a and $g : \Omega' \to \mathbb{C}$ is holomorphic at $f(a)$, then $g \circ f : \Omega \to \mathbb{C}$ is also holomorphic at a and $(g \circ f)'(a) = g'(f(a))f'(a)$. Just as in the real case this follows from items 1 and 2 above.

We shall say a holomorphic function is *regular* on Ω if $f'(a) \neq 0$ for every $a \in \Omega$.

Corollary 1.2.2 *If f is locally invertible at a, then f is regular there.*

Proof. Let g be the holomorphic local inverse to f at a. Then $g(f(z)) = z$ for z in some neighborhood of a. As we saw the derivative of z is 1, differentiating, we get $g'(f(z))f'(z) = 1$. In particular, this holds at $z = a$. Therefore, $f'(a) \neq 0$. □

In Section 1.4 we shall see why the g above is holomorphic and that the converse of this corollary is also true.

1.3 Some important examples

Here, as in the previous section, we shall rely on various useful facts of real analysis.

We now define what is, taken in its most general form, the most important function in mathematics, namely the exponential function written e^z or $\exp(z)$. e^z is defined by the power series

$$e^z = \sum_{n=0}^{\infty} \frac{z^n}{n!}, z \in \mathbb{C}.$$

This series converges absolutely and uniformly on compacta of $\mathbb{C}$. To check absolute convergence note that $|z^n| = |z|^n$, for every n. Hence $\sum_{n=0}^{\infty} \frac{|z^n|}{n!} = e^{|z|}$, the real exponential function, which converges absolutely. Also, if $|z| \leq c$, then the tail $|\sum_{n=p}^{q} \frac{z^n}{n!}|$ can be estimated by $\sum_{n=p}^{q} \frac{|z|^n}{n!} \leq \sum_{n=p}^{q} \frac{c^n}{n!}$ which is the tail of e^c, $c \in \mathbb{R}$. Since this converges as p and q tend to infinity by completeness the series for e^z also converges uniformly on compacta. As we shall see in Section 1.7 these results are actually quite general.

Corollary 1.3.1 $e^{z+w} = e^z e^w$.

Proof. Now $e^{z+w} = \sum_{n=0}^{\infty} \frac{(z+w)^n}{n!}$. By the binomial theorem, $(z+w)^n = \sum_{k=0}^{n} \frac{n!}{k!(n-k)!} z^k w^{n-k}$. Hence, $e^{z+w} = \sum_{n=0}^{\infty} \sum_{k=0}^{n} \frac{1}{k!(n-k)!} z^k w^{n-k}$. By absolute convergence we can rearrange the order of summation in this series. Letting $m = n - k$ we get $\sum_{n=0}^{\infty} \sum_{m=0}^{\infty} \frac{z^n}{n!} \frac{w^m}{m!}$. On the other hand, also by rearrangement this is $\sum_{n=0}^{\infty} \frac{z^n}{n!} \sum_{m=0}^{\infty} \frac{w^m}{m!}$, which is just $e^z e^w$. □

Corollary 1.3.2 *For all $t \in \mathbb{R}$ and $z \in \mathbb{C}$, $|e^{tz}| = |e^z|^t$.*

Proof. For all z and w, by Corollary 1.3.1, $e^z e^w = e^{z+w}$. Therefore, $e^{nz} = (e^z)^n$, for $z \in \mathbb{C}$ and n a positive integer. But also $e^z e^{-z} = e^0 = 1$ so $(e^z)^{-1} = e^{-z}$. Hence for n a positive integer

$$e^{-nz} = e^{n(-z)} = e^{-zn} = ((e^z)^{-1})^n = (e^z)^{-n}.$$

Thus the conclusion holds for all $z \in \mathbb{C}$ and $n \in \mathbb{Z}$.

Next let $r = \frac{p}{q}$ be rational. Then $(e^{\frac{p}{q}z})^q = e^{q\frac{p}{q}z} = e^{pz} = (e^z)^p$. Hence $|e^{\frac{p}{q}z}|^q = |e^z|^p$ and so $|e^{\frac{p}{q}z}| = |e^z|^{\frac{p}{q}}$. This shows $|e^{rz}| = |e^z|^r$ for all $z \in \mathbb{C}$ and $r \in \mathbb{Q}$.

Since for fixed $z \in \mathbb{C}$ both sides of the equation in the statement of the Lemma are continuous functions of t and $\mathbb{Q}$ is dense in $\mathbb{R}$, it holds for all $t \in \mathbb{R}$. □

Exercise 1.2 *Show* $\lim_{n\to\infty}(1 + \frac{z}{n})^n = e^z$.

Now similar reasoning to that used to define exp shows that the power series defining sin and cos in the real domain work just as well in the complex domain. Thus we get the functions

$$\sin z = \sum_{n=0}^{\infty}(-1)^n \frac{z^{2n+1}}{(2n+1)!}$$

and

$$\cos z = \sum_{n=0}^{\infty}(-1)^n \frac{z^{2n}}{(2n)!},$$

all for $z \in \mathbb{C}$.

Corollary 1.3.3 $e^{iz} = \cos z + i \sin z$.

To verify this, write the power series for

$$e^{iz} = \sum_{n=0}^{\infty} \frac{(iz)^n}{n!} = 1 + iz - \frac{z^2}{2!} - i\frac{z^3}{3!} + \frac{z^4}{4!} + \cdots.$$

As we have seen, this series converges absolutely and so can be rearranged. Combining the even and the odd terms we get the relation of the corollary.

In particular, taking z real in that relation we get Euler's relation. (Sometimes Corollary 1.3.3 itself is called Euler's relation).

$$e^{i\theta} = \cos\theta + i\sin\theta\,.$$

Thus, $e^{i\theta}$ lies on the unit circle. In fact, it parameterizes the unit circle (modulo 2π).

An instance of this formula is $e^{2\pi i} = 1$. Combining this fact with Corollary 1.3.1 tells us for every integer n, $e^{z+2\pi in} = e^z$. Thus e^z is periodic of period $2\pi i$. If e^z assumes a value, it must assume that value infinitely often. In particular, unlike the real situation, e^z is not invertible.

Another conclusion to be drawn from all this is if $z = x + iy \in \mathbb{C}$, then $|e^z| = |e^{x+iy}| = |e^x e^{iy}| = |e^x||e^{iy}|$. Since $e^x > 0$ and $|e^{iy}| = 1$ we see

$$|e^z| = e^{\Re z}\,.$$

In particular, e^z is never zero. In fact, from Corollary 1.3.1 it follows that $(e^z)^{-1} = e^{-z}$. Also notice that, conversely, if $e^z = 1$ for some z, then $e^x e^{iy} = 1 \cdot 1$. By the uniqueness of the polar decomposition $e^x = 1$ and $e^{iy} = 1$, so $x = 0$ and y is congruent to zero modulo 2π and therefore z is congruent to zero modulo $2\pi i\mathbb{Z}$.

We now come to DeMoivre's theorem. Namely for z, $w \in \mathbb{C}^\times$,

$$|zw| = |z||w|\,,$$

which we already know and

$$\arg zw = \arg z + \arg w \bmod(2\pi)\,.$$

This is because $z = |z|\frac{z}{|z|} = |z|e^{i\theta}$ and similarly $w = |w|e^{i\phi}$. Therefore,

$$zw = |z||w|e^{i\theta}e^{i\phi} = |z||w|e^{i(\theta+\phi)}\,.$$

By uniqueness of the polar decomposition, both relations follow.

Exercise 1.3 *Let z be a non-zero complex number and $n \geq 2$ an integer. Show that z has n distinct n^{th} roots. Taking $n \geq 3$, draw a diagram of these n distinct roots and show that they lie on a regular polygon with n sides.*

From the calculation above we see that the range of e^z is all of $\mathbb{C}^\times$. For if $w \neq 0$, then $w = |w|e^{i\phi}$ where $|w| > 0$. Therefore, $|w| = e^t$ for some real t, and so $w = e^t e^{i\phi} = e^{t+i\phi}$.

A natural question is, when does $e^z = e^{z_1}$? Since $(e^{z_1})^{-1} = e^{-z_1}$, this occurs if and only if $e^z e^{-z_1} = 1$. That is, iff $e^{z-z_1} = 1$. But then, as we saw, z is congruent to z_1 modulo $2\pi i\mathbb{Z}$. In particular, this shows e^z is locally invertible because if $z \in \mathbb{C}$ and we take an open disk D about z of radius 2π, or even an open horizontal strip of width 2π, then on this domain e^z is 1:1. Later we will construct holomorphic functions which locally invert e^z.

Proposition 1.3.4 *e^z is an entire function. Its derivative is itself.*

Proof. First we note that, just as in calculus, by the change of variable $h = z - a$ to see f is holomorphic at $a \in \mathbb{C}$, it is equivalent to check that

$$\lim_{h \to 0} \frac{f(a+h) - f(a)}{h}$$

exists. Here $e^{a+h} = e^a e^h$ so that

$$\lim_{h \to 0} \frac{e^{a+h} - e^a}{h} = e^a \lim_{h \to 0} \frac{e^h - 1}{h}.$$

Thus e^z is entire if it is holomorphic at 0. We will show

$$\lim_{h \to 0} \frac{e^h - 1}{h} = 1.$$

Hence e^z is an entire function and its derivative is again e^z. Since $e^h = \sum_{n=0}^{\infty} \frac{h^n}{n!}$, we get $\frac{e^h - 1}{h} = 1 + \frac{h}{2!} + \frac{h^2}{3!} \dots$. Since this series also converges uniformly on compacta and the uniform limit on compacta of continuous functions is continuous (the partial sums being polynomials and therefore continuous), we see that this series gives a continuous function. Therefore its limiting value at $h = 0$ is gotten by evaluating at 0. Here we get 1. □

Since for all z, $e^{iz} = \cos z + i \sin z$, we also get $e^{-iz} = \cos(-z) + i\sin(-z)$. On the other hand inspection of the series defining sin and cos make clear that sin is an odd function and cos an even function. Thus $\sin(-z) = -\sin z$ and $\cos(-z) = \cos z$. Substituting into the above gives $e^{-iz} = \cos z - i \sin z$. Adding to and subtracting this from $e^{iz} = \cos z + i \sin z$ tells us that

$$\cos z = \frac{1}{2}(e^{iz} + e^{-iz})$$

and

$$\sin z = \frac{1}{2i}(e^{iz} - e^{-iz}).$$

Hence, we can also calculate the derivatives of these functions, getting

$$\frac{d}{dz} \sin z = \frac{1}{2i}(ie^{iz} + ie^{-iz}) = \frac{1}{2}(e^{iz} + e^{-iz}) = \cos z$$

and similarly, $\frac{d}{dz} \cos z = -\sin z$. Hence, these are entire functions.

We remark that since $\frac{d}{dz} \sin z = \cos z$ everywhere, taking $z = 0$ tells us since $\sin 0 = 0$ and $\cos 0 = 1$ that $\lim_{z \to 0} \frac{\sin z}{z} = 1$.

This statement is slightly stronger than the corresponding one for real functions because z can approach zero in more essentially different ways. Turning to the hyperbolic functions, $\sinh z$ and $\cosh z$, these can also be defined by extending the usual real power series into the complex domain.

$$\sinh z = \sum_{n=0}^{\infty} \frac{z^{2n+1}}{(2n+1)!}$$

and

$$\cosh z = \sum_{n=0}^{\infty} \frac{z^{2n}}{(2n)!}.$$

Similarly to sin and cos, it is easily checked that $\sinh z = \frac{e^z - e^{-z}}{2}$ and $\cosh z = \frac{e^z + e^{-z}}{2}$. Exactly as above, from this we see $\frac{d}{dz} \sinh z = \cosh z$ and $\frac{d}{dz} \cosh z = \sinh z$. Hence, these are also entire functions.

Definition 1.3.5 We say a complex valued function defined on a domain is *analytic* if it is represented by a convergent power series. Similarly, if a real valued function of a real variable is represented by a convergent power series we call it a *real analytic* function.

We conclude this section with two remarks. First, all these differentiation results could also have been gotten by termwise differentiation of the appropriate power series. However, proof of this fact would have required further knowledge of power series than we presently have. In Chapter 3 we will see why this is so. Secondly, it is obvious that our process of extending real analytic functions defined on an interval, a half line, or the whole real axis by convergent power series to the corresponding complex domain is quite general and has nothing to do with the particulars of the functions we have been considering in this section (Figure 3.6). So, for example, if $f(x) = \sum_{n=0}^{\infty} a_n(x-a)^n$ converges for $|x-a| < r$, where the x, a_n and a are real and $0 < r \leq \infty$, then we get a complex power series $f(z) = \sum_{n=0}^{\infty} a_n(z-a)^n$ which converges for all $|z-a| < r$, where $0 < r \leq \infty$.

Exercise 1.4 *Use the functional equation for e^z to prove the addition formulas for* sin, cos, sinh *and* cosh.

1.4 The Cauchy-Riemann equations

Let $f : \Omega \to \mathbb{C}$ be a map which is holomorphic at a point a. If $z = x + iy \in \Omega$, write $f(z) = u(x,y) + iv(x,y)$. We know $\lim_{z\to a} \frac{f(z)-f(a)}{z-a} = f'(a)$ exists no matter how $z \to a$. We shall compute this limit in two different ways. First, let $z \to a$ by keeping x constant and y varying and then by keeping y constant and x varying. Let $a = b + ic$ and $z = x + ic$, where $x \to b$. Then, $z - a = x - b$ so $z \to a$ and

$$\frac{f(z)-f(a)}{z-a} = \frac{u(x,c)-u(b,c)}{x-b} + i\frac{v(x,c)-v(b,c)}{x-b}.$$

Taking limits as $z \to a$ we get

$$f'(a) = \frac{\partial u}{\partial x}(b,c) + i\frac{\partial v}{\partial x}(b,c).$$

On the other hand if $z = b + iy$ where $y \to c$, then $z - a = i(y - c)$; so z also approaches a and

$$\frac{f(z) - f(a)}{z - a} = \frac{u(b, y) - u(b, c)}{i(y - c)} + i\frac{v(b, y) - v(b, c)}{i(y - c)}.$$

Since $i^{-1} = -i$, taking limits this time we get

$$f'(a) = -i\frac{\partial u}{\partial y}(b, c) + \frac{\partial v}{\partial y}(b, c).$$

Thus $\frac{\partial u}{\partial x}(a) = \frac{\partial v}{\partial y}(a)$ and $\frac{\partial v}{\partial x}(a) = -\frac{\partial u}{\partial y}(a)$. These are called the Cauchy-Riemann equations at a.

In general, when $f : \Omega \to \mathbb{R}^2$ is a smooth function, the Jacobian matrix $J_f(a)$ has the form

$$\begin{pmatrix} \frac{\partial u}{\partial x}(a) & \frac{\partial u}{\partial y}(a) \\ \frac{\partial v}{\partial x}(a) & \frac{\partial v}{\partial y}(a) \end{pmatrix}$$

However, because of the Cauchy-Riemann equations, in the holomorphic case we have:

Corollary 1.4.1 *If f is holomorphic at a, then $J_f(a)$ has the form*

$$\begin{pmatrix} \alpha & \beta \\ -\beta & \alpha \end{pmatrix}.$$

Corollary 1.4.2 *If f is holomorphic at a, then $|f'(a)|^2 = \det J_f(a)$. In particular, f is regular at a if and only if f is locally invertible there.*

Proof. As we saw, $f'(a) = \frac{\partial u}{\partial x}(a) + i\frac{\partial v}{\partial x}(a)$. Hence $|f'(a)|^2 = \frac{\partial u}{\partial x}(a)^2 + \frac{\partial v}{\partial x}(a)^2$. But by the Cauchy-Riemann equations, this is exactly $\det J_f(a)$. In particular, f is regular iff $J_f(a)$ is invertible. By the real inverse function theorem this is equivalent to f being locally invertible as a real mapping. □

Corollary 1.4.3 *If f is holomorphic on Ω, then f is C^1 and $\frac{\partial u}{\partial x} = \frac{\partial v}{\partial y}$, $\frac{\partial v}{\partial x} = -\frac{\partial u}{\partial y}$. Also, $|f'(z)|^2 = \det J_f(z)$. In particular, f is regular on Ω if and only if f is locally invertible everywhere on Ω.*

Corollary 1.4.3 gives us a way of testing whether or not a complex function is holomorphic. For example, conjugation is not holomorphic since $u_x = 1$, but $v_y = -1$.

A real valued function u defined on a domain Ω is called *harmonic* if $u_{xx} + u_{yy} = 0$ everywhere on the domain. The operator $\Delta = \frac{\partial^2}{\partial x^2} + \frac{\partial^2}{\partial y^2}$ is called the Laplacian.

Corollary 1.4.4 *If f is holomorphic on Ω and f is C^2, then its real and imaginary parts are harmonic.*

We remark that in Chapter 3 we shall see f must actually be C^∞.

Proof. By the Cauchy-Riemann equations $u_x = v_y$ and $v_x = -u_y$. Differentiating we get $u_{xx} = v_{xy}$ and $u_{yy} = -v_{yx}$. But since v (and u) are both C^2, we know $v_{yx} = v_{xy}$. Therefore $u_{xx} + u_{yy} = 0$. Similarly, $v_{xx} + v_{yy} = 0$. □

We now turn to the converse of the Cauchy-Riemann equations.

Theorem 1.4.5 *Let $f : \Omega \to \mathbb{C}$ be a map, $f(z) = u(x,y) + iv(x,y)$, where u and v are C^1 functions. If the Cauchy-Riemann equations are satisfied at a point a, then f is holomorphic at a.*

Proof. By the linear approximation theorem for real valued functions of several real variables we see that

$$u(z) - u(a) = u_x(a)(x-b) + u_y(a)(y-c) + \epsilon_1(z)(x-b) + \epsilon_2(z)(y-c)$$

and

$$v(z) - v(a) = v_x(a)(x-b) + v_y(a)(y-c) + \epsilon_3(z)(x-b) + \epsilon_4(z)(y-c),$$

where, as above, $a = (b,c)$ and each ϵ_j tends to zero as $z \to a$. Therefore,

$$\begin{aligned} f(z) - f(a) = {} & u_x(a)(x-b) + u_y(a)(y-c) + \epsilon_1(z)(x-b) \\ & + \epsilon_2(z)(y-c) + i(v_x(a)(x-b) + v_y(a)(y-c)) \\ & + \epsilon_3(z)(x-b) + \epsilon_4(z)(y-c)) \,. \end{aligned}$$

That is,

$$\frac{f(z)-f(a)}{z-a} = (u_x(a)+iv_x(a))\frac{x-b}{z-a} - i(u_y(a)+iv_y(a))i\frac{y-c}{z-a}$$
$$+ (\epsilon_1(z)+i\epsilon_3(z))\frac{x-b}{z-a} + (\epsilon_2(z)+i\epsilon_4(z))\frac{y-c}{z-a}.$$

By the Cauchy-Riemann equations,

$$\frac{f(z)-f(a)}{z-a} = (u_x(a)+iv_x(a))\frac{z-a}{z-a} + (\epsilon_1(z)+i\epsilon_3(z))\frac{x-b}{z-a}$$
$$+ (\epsilon_2(z)+i\epsilon_4(z))\frac{y-c}{z-a}.$$

Taking into account that each ϵ_j tends to zero and $|\frac{x-b}{z-a}|$ and $|\frac{y-c}{z-a}| \leq 1$, we see that

$$\lim_{z\to a}\frac{f(z)-f(a)}{z-a} = u_x(a)+iv_x(a).$$

□

Corollary 1.4.6 *Let $f:\Omega\to\mathbb{C}$ be a map, where $\Re f$ and $\Im f$ are C^1 functions. If f satisfies the Cauchy-Riemann equations on Ω, then f is holomorphic.*

Corollary 1.4.7 *Let $f:\Omega\to\mathbb{C}$ be a regular holomorphic map. Then local inverses of f are holomorphic.*

Proof. Since f is regular, we know f is locally invertible by a smooth real function. The only question is whether the local inverse is holomorphic. But we know that for each $z\in\Omega$, $J_f(z)$ has the form

$$\begin{pmatrix} \alpha & \beta \\ -\beta & \alpha \end{pmatrix}.$$

As is well known from linear algebra, $J_f(z)^{-1}$ has the form

$$\begin{pmatrix} \frac{\alpha}{\alpha^2+\beta^2} & -\frac{\beta}{\alpha^2+\beta^2} \\ \frac{\beta}{\alpha^2+\beta^2} & \frac{\alpha}{\alpha^2+\beta^2} \end{pmatrix}.$$

Since by the chain rule the tangent mapping of f^{-1} is $J_f(z)^{-1}$, it follows from the result immediately above that f^{-1} is holomorphic. $\square$

Because even for real functions a local diffeomorphism is open, we get the following:

Corollary 1.4.8 *Let $f : \Omega \to \mathbb{C}$ be a regular holomorphic map. Then f is an open map. In particular, $f(\Omega)$ is open.*

Exercise 1.5 *Let f be a holomorphic function in a domain, Ω and Δ be the Laplacian. Show that*

$$\Delta(|f(z)|^2) = 4|f'(z)|^2.$$

Hence if $\{f_1, \ldots, f_k\}$ are holomorphic functions and $\sum_{j=1}^{k} |f_j(z)|^2$ is harmonic, each f_j is constant.

1.5 Some elementary differential equations

From this point on we shall require that domains Ω be connected open sets and we shall assume this is understood without it being explicitly mentioned in the sequel.

We observe that if Ω is merely open, then each connected component is a domain. Also notice that if γ is a continuous curve in an open set that starts out at say a, then it remains in the component of Ω containing a throughout its trajectory. Thus, there is no harm that can come from this restriction. The reader is invited to prove these statements.

Lemma 1.5.1 *Let $f : D \to \mathbb{C}$ be a holomorphic map, where $D(a, r)$ is the disk centered at a of radius $r > 0$. If $f' \equiv 0$, then f is constant.*

Proof. Let $z \in D$. By convexity, the line segment $\gamma(t) = ta + (1-t)z$ lies in D for $0 \leq t \leq 1$. This is because $|\gamma(t) - a| = |(t-1)a + (1-t)z| = |1-t||z-a| \leq |z-a| < r$. Now as we saw, $\frac{d}{dt}f(\gamma(t)) = f'(\gamma(t))\gamma'(t)$. Since $f' \equiv 0$ we see $\frac{d}{dt}f(\gamma(t)) = 0$ for all t. Applying the mean value theorem to the components f_j of $f(\gamma(t))$, we conclude $f_j(a) - f_j(z) = f_j(\gamma(1)) - f_j(\gamma(0)) = 0$ for each j. Hence $f(z) = f(a)$, a constant. $\square$

Proposition 1.5.2 *Let $f : \Omega \to \mathbb{C}$ be a holomorphic map. If $f' \equiv 0$, then f is constant.*

Proof. Let z_0 be fixed and z a variable point in Ω. Since Ω is connected and open, it is arcwise connected. That is, these two points can be joined by a continuous arc $\gamma : [a, b] \to \Omega$. By compactness of $[a, b]$ and continuity of γ we know the trajectory, $\gamma([a, b])$, is compact. At each point $\gamma(t)$ of the trajectory choose a small disk D_t centered at $\gamma(t)$ and contained in Ω. This covering of the trajectory has a finite subcover, say $D_{t_1}, \ldots, D_{t_n}$. By Lemma 1.5.1, f is constant on each of these disks. But each successive pair of these disks must overlap or together they could not cover the trajectory. Now the union of two connected sets with a point in common must itself be connected so $D_{t_j} \cup D_{t_{j+1}}$ is connected. Hence, so is its image under f. Therefore, f is constant on the union. It follows that f is constant on the trajectory of γ and in particular, $f(z_0) = f(\gamma(a)) = f(\gamma(b)) = f(z)$. □

Exercise 1.6 1. *Show if the domain were not connected this result would be false.*

2. *Prove that a connected space such as Ω is arcwise connected.*

Corollary 1.5.3 *Let f, $g : \Omega \to \mathbb{C}$ be holomorphic maps. If $f' = g'$, then $f - g$ is constant.*

This is clear since $(f - g)' = f' - g' = 0$. Therefore $f - g$ is constant.

Corollary 1.5.4 *f is a polynomial if and only if $f^{(n)} = 0$ for some n.*

Proof. Here we take $\Omega = \mathbb{C}$. If f is a polynomial of degree $n - 1$, then as was mentioned earlier, $f^{(n)} = 0$. Conversely, suppose $f^{(n)} = 0$ for some n. Since $f^{(n-1)'} = 0$, $f^{(n-1)}$ is constant, say c. Let $g = f^{(n-2)}$ and $h(z) = cz$. Since the derivatives of g and h are equal, $g - h = b$, a constant, so $f^{(n-2)} = cz + b$. Continuing in this way by induction, we see that f is a polynomial of degree at most $n - 1$. □

Corollary 1.5.5 *Let $f : \Omega \to \mathbb{C}$ be a holomorphic map and γ be a smooth curve in Ω. If $f(\Omega)$ is contained in the trajectory of γ, then f*

must be constant. In particular, if f takes only real, or purely imaginary values, or if $|f|$ is constant, then f must itself be constant.

Proof. Since the image of the domain is a curve and so of lower dimension, $J_f(z)$ must be singular at every point. As we know, $\det J_f(z) = |f'(z)|^2$. Hence $f'(z) = 0$ for all $z \in \Omega$. By Proposition 1.5.2, f must be constant. □

Later we shall see that this result also follows from the area theorem of Section 3.4.

Corollary 1.5.6 *Let $f : \Omega \to \mathbb{C}$ be a holomorphic map. Then the following are equivalent.*

(i) *f is constant.*
(ii) *$\Re(f)$ and $\Im(f)$ are constant.*
(iii) *$f' = 0$.*
(iv) *$|f|$ is constant.*

We already see that (i), (ii) and (iii) are all equivalent and they imply (iv). By Corollary 1.5.5, (iv) also implies (i).

1.6 Conformality

The definition of conformality actually has little to do with dimension 2 or with complex analysis for that matter. Let Ω be a domain in $\mathbb{R}^n$ and f a real C^∞ map $\Omega \to \mathbb{R}^n$ which is locally invertible.

Definition 1.6.1 We say f is *conformal at a point* $a \in \Omega$ if for each pair of smooth curves γ and δ in Ω passing through a, say at $t = t_0$, the angle between their tangent vectors $\gamma'(t_0)$ and $\delta'(t_0)$ equals the angle between the tangent vectors $(f \circ \gamma)'(t_0)$ and $(f \circ \delta)'(t_0)$, their images under f. If f is conformal at every point in Ω, we just say f is *conformal.*

Our purpose here is to produce a large number of conformal mappings in $\mathbb{R}^2 = \mathbb{C}$ and actually to characterize conformal maps in $\mathbb{R}^2$. In this connection a useful application of the chain rule is the following:

Corollary 1.6.2 *Let $f : \Omega \to \Omega'$ be a holomorphic function between domains and γ be a smooth curve in Ω. Then $f \circ \gamma$ is a smooth curve in Ω' and $f \circ \gamma'(t_0)$, its tangent vector at t_0, is $f'(\gamma(t_0))\gamma'(t_0)$.*

The main result here is the following:

Theorem 1.6.3 *Let $f : \Omega \to \mathbb{C}$ be a regular holomorphic map. Then f is conformal. Conversely, if $f : \Omega \to \mathbb{C}$ is conformal, then f is holomorphic and regular on Ω.*

Proof. Suppose f is holomorphic at a, and γ and δ are smooth curves in Ω passing through a at t_0. By the chain rule,

$$(f \circ \gamma)'(t_0) = f'(\gamma(t_0))\gamma'(t_0)$$

and

$$(f \circ \delta)'(t_0) = f'(\delta(t_0))\delta'(t_0)\,.$$

Hence,

$$\arg(f \circ \gamma)'(t_0) = \arg(f'(\gamma(t_0))) + \arg(\gamma'(t_0))$$

and similarly

$$\arg(f \circ \delta)'(t_0) = \arg(f'(\delta(t_0))) + \arg(\delta'(t_0))\,.$$

Since these curves both pass through a at t_0 it follows that

$$\arg(f'(\gamma(t_0))) = \arg(f'(\delta(t_0)))\,.$$

So we get

$$\arg(f \circ \gamma)'(t_0) - \arg(f \circ \delta)'(t_0) = \arg(\gamma'(t_0)) - \arg(\delta'(t_0))\,.$$

This means the angle between $\gamma'(t_0)$ and $\delta'(t_0)$ equals the angle between $(f \circ \gamma)'(t_0)$ and $(f \circ \delta)'(t_0)$. Since the smooth curves are arbitrary, as is $a \in \Omega$, f is conformal.

Conversely, suppose f is conformal. Linear algebra tells us that the tangent mapping $J_f(a)$ at a is given by $J_f(a) = \lambda R_\theta$, where $\lambda > 0$ and R_θ is rotation by angle θ. Therefore, the Cauchy-Riemann equations are satisfied at each point. Since f is a C^∞ map, by Corollary 1.4.6 it is holomorphic at a. Because f is locally invertible, by Corollary 1.2.2, f is also regular at a. □

Exercise 1.7 *Show that, as above, an angle preserving linear map T of the plane (or indeed of $\mathbb{R}^n$) must be of the form a stretch followed by a rotation. (Notice that a stretch commutes with any other linear transformation).*

(Suggestion: First show that $\det T > 0$. Then consider $S = \frac{1}{(\det T)^{\frac{1}{n}}} T$. Show that S also preserves angles. Prove $\det S = 1$ and therefore S also preserves area (volume). Because S preserves both angles and area it must preserve congruences and since $\det S = 1$, it must actually be a rotation. Hence T is a positive multiple of a rotation. These are the infinitesimal conformal maps in $\mathbb{R}^n$.)

1.7 Power series

In this section we very briefly give the main elementary results on power series, usually referred to as Abel's lemma. To do so we first deal with the geometric series.

Proposition 1.7.1 *For $|z| < 1$ and $b \in \mathbb{C}$ the series*

$$\sum_{n=0}^{\infty} bz^n = \frac{b}{1-z}.$$

Proof. We may clearly assume $b = 1$. Then $z\sum_{n=0}^{N} z^n = \sum_{n=1}^{N+1} z^n$. Therefore the difference $\sum_{n=0}^{N} z^n - z\sum_{n=0}^{N} z^n = 1 - z^{N+1}$. Since $|z| < 1$, it follows easily that z^n tends to zero and so $1 - z^{N+1}$ tends to 1. This means that $(1-z)\sum_{n=0}^{N} z^n$ tends to 1 and since $z \neq 1$ that $\sum_{n=0}^{N} z^n$ tends to $\frac{1}{1-z}$. □

Definition 1.7.2 In all that follows, if f is a bounded complex valued function on a space X we shall denote its sup norm over X by $\| f \|_X$.

Proposition 1.7.3 *For $n \geq 0$ let M_n be a sequence of positive numbers such that $\sum_{n=0}^{\infty} M_n$ is convergent. Suppose $\sum_{n=0}^{\infty} f_n(x)$ is a series of complex valued functions defined on a metric space X and for all n, $\| f_n \| \leq M_n$. Then $\sum_{n=0}^{\infty} f_n(x)$ is uniformly convergent on X.*

Proof. Since $\sum_{n=0}^{\infty} f_n(x) \leq \sum_{n=0}^{\infty} M_n$ and the latter is convergent evidently $\sum_{n=0}^{\infty} f_n(x)$ is convergent to $f(x)$ at every point $x \in X$. To see that the convergence is uniform we apply the Cauchy criterion. For all x,

$$\left| f(x) - \sum_{n=0}^{N} f_n(x) \right| = \left| \sum_{n=N+1}^{\infty} f_n(x) \right| \leq \sum_{n=N+1}^{\infty} |f_n(x)| \leq \sum_{n=N+1}^{\infty} M_n .$$

But since $\sum_{n=0}^{\infty} M_n$ is convergent its tail $\sum_{n=N+1}^{\infty} M_n$ tends to zero as $N \to \infty$. Hence if $\epsilon > 0$ is given, then for all $x \in X$. $|f(x) - \sum_{n=0}^{N} f_n(x)| < \epsilon$. Hence for N large enough $\| f - \sum_{n=0}^{N} f_n \|_X \leq \epsilon$. □

Lemma 1.7.4 *If the power series $\sum_{n=0}^{\infty} a_n(z_0 - a)^n$ is convergent for some z_0, then it converges absolutely for all $|z - a| < |z_0 - a|$.*

Proof. Because $\sum_{n=0}^{\infty} a_n(z_0 - a)^n$ is convergent its nth term tends to zero. So, given $B > 0$, eventually $|a_n||z_0 - a|^n < B$. By taking B larger we can absorb the finite number of earlier terms and get $|a_n||z_0 - a|^n < B$ for all n. Now

$$\sum_{n=0}^{\infty} |a_n|(|z - a|)^n = \sum_{n=0}^{\infty} |a_n| \frac{(|z - a|)^n}{|z_0 - a|^n} |z_0 - a|^n \leq B \sum_{n=0}^{\infty} \left| \frac{z - a}{z_0 - a} \right|^n .$$

But since $|z - a| < |z_0 - a|$, $|\frac{z-a}{z_0-a}| < 1$ and so we have a convergent geometric series. This means what is dominated by it must also converge. □

Theorem 1.7.5 *Let $f(z) = \sum_{n=0}^{\infty} a_n(z - a)^n$ be a convergent power series for $|z - a| < r_0$, where $0 < r_0 \leq \infty$. Then this series is absolutely and uniformly convergent for $|z - a| \leq r$, where $r < r_0$.*

Proof. Since $r < r_0$, the absolute convergence for $|z - a| \leq r$ follows from Abel's lemma. Therefore, $\sum_{n=0}^{\infty} |a_n| r^n < \infty$. Taking $M_n = |a_n| r^n$ for all $n \geq 0$ we see that $|a_n(z - a)^n| \leq M_n$ for all n and all $|z - a| \leq r$. Since $\sum_{n=0}^{\infty} M_n$ converges, it follows from the Weierstrass M-test that $\sum_{n=0}^{\infty} a_n(z - a)^n$ converges uniformly for $|z - a| \leq r$. □

Chapter 2

Integration Along a Contour

2.1 Curves and their trajectories

In this section we work out the basic facts of integral calculus in the complex domain. Although we have already met the notion of a contour, here we will formalize it and single out those contours we shall be using throughout the sequel.

Definition 2.1.1 A continuous map $\gamma : [a, b] \to \Omega$, where Ω is a domain in $\mathbb{C}$, will be called a *continuous curve* or *arc* in Ω. We call $[a, b]$ the *parameter domain*, $\gamma(a)$ and $\gamma(b)$ the *endpoints*, and $\gamma[a, b]$ the *trajectory*. We shall say γ is *smooth* if it is a C^∞ curve.

If $\gamma(a) = \gamma(b)$, we say the curve is closed. If γ is closed and $\gamma(t_1) \neq \gamma(t_2)$, for $t_1 \neq t_2$, where t_1, $t_2 \in (a, b)$, we shall say γ is a *simple* closed curve.

We shall largely be interested in simple closed curves . In this case it is not so important to distinguish between the curve and its trajectory. In general, when the significance is clear, we may not always distinguish these two notions.

A very special case of the Jordan Schoenflies theorem [6, p. 68], tells us if γ is a piecewise smooth simple closed curve in $\mathbb{R}^2$, then its complement has two components, one bounded (the interior) and the other unbounded (the exterior). This is called the Jordan curve theorem. Moreover, there is a homeomorphism of the plane taking the curve to the unit circle and the interior of the curve to the interior of the unit circle. In particular, the interior of γ is connected and simply connected. (Perhaps it should also be remarked that the analogue of the Jordan Schoenflies theorem in three dimensional space is false).

Here are four examples of curves all of which have the same domain. These curves are in $\Omega = \mathbb{C}$, $a \in \mathbb{C}$ and $r > 0$.

1. $\gamma_1(t) = re^{2\pi it} + a$, $0 \le t \le 1$.

2. $\gamma_2(t) = re^{2\pi it^2} + a$, $0 \le t \le 1$.

3. $\gamma_3(t) = re^{4\pi it} + a$, $0 \le t \le 1$.

4. $\gamma_4(t) = re^{-2\pi it} + a$, $0 \le t \le 1$.

Each of these curves is a C^∞ closed curve. They all have the same trajectory, namely the circle centered at a of radius r. γ_1, γ_2 and γ_4 are simple; γ_3 is not.

Now suppose γ is smooth. This means the component functions of $\gamma(t) = x(t) + iy(t)$ are smooth. When we have a smooth curve, its derivative $\gamma'(t) = x'(t) + iy'(t)$ is called the tangent vector (or velocity vector) at t. Its length, $|\gamma'(t)|$, is called the speed at t. So for example, $\gamma_1'(t) = 2\pi ire^{it}$ and its speed is a constant, $2\pi r$. It traces out the circle in a counter-clockwise direction. On the other hand γ_4 traces out the circle at the same constant speed, but in a clockwise direction. The speed of γ_2 is not constant.

Definition 2.1.2 When a simple closed curve goes around in a counter-clockwise direction, we say it is positively oriented. Otherwise, we say it is negatively oriented.

As we saw earlier, another example of a curve, this time not closed, is the straight line joining the distinct points α and β. Here $\gamma(t) = (1-t)\alpha + t\beta$, where $0 \le t \le 1$. The tangent vector is constant; $\gamma'(t) = \beta - \alpha$.

We shall be especially interested in the so-called piecewise smooth curves (respectively piecewise linear curves). These are curves where the parameter domain has a finite number of special points in between which the curve is smooth (respectively linear), but at which the curve may not be smooth. However, at least it is continuous at these points. Clearly such piecewise smooth curves include piecewise linear ones.

If γ is a smooth curve we define its length by

$$L(\gamma) = \int_a^b |\gamma'(t)|dt\,.$$

Or, in infinitesimal form $ds = \sqrt{x'(t)^2 + y'(t)^2}dt$. For a piecewise smooth curve, γ, its length is the sum of the lengths of the finite number of smooth pieces. Alternatively, since Riemann integration is unaffected by a finite number of such (jump) discontinuities, here also $L(\gamma) = \int_a^b |\gamma'(t)|dt$.

If γ is a continuous curve with parameter domain $[a, b]$, we define $\int_a^b \gamma dt$ component-wise. Namely, $\int_a^b \gamma dt = \int_a^b \Re(\gamma)dt + i\int_a^b \Im(\gamma)dt$. Clearly this operation is linear. If δ is another continuous curve with the same parameter domain we have

$$\int_a^b (\gamma + \delta)dt = \int_a^b \gamma dt + \int_a^b \delta dt$$

and

$$\int_a^b c\gamma dt = c\int_a^b \gamma dt\,.$$

Lemma 2.1.3 *If δ is a continuous curve defined on $[a, b]$, then*

$$\left|\int_a^b \delta dt\right| \leq \int_a^b |\delta| dt\,.$$

Proof. If $|\int_a^b \delta dt| = 0$ the inequality is surely true. Otherwise, choose a real θ so that $e^{i\theta}\int_a^b \delta dt = |\int_a^b \delta dt|$. Hence, $|\int_a^b \delta dt| = \int_a^b e^{i\theta}\delta dt$. So $g(t) = e^{i\theta}\delta(t)$ is a function whose integral is real (and positive). That is, $\int_a^b g(t)dt = \int_a^b \Re(g(t))dt + i\int_a^b \Im(g(t))dt$ is real. This means that $\int_a^b \Im(g(t))dt = 0$ and $\int_a^b e^{i\theta}\delta(t)dt = \int_a^b \Re(g(t))dt > 0$. Hence,

$$\left|\int_a^b g(t)dt\right| = \int_a^b e^{i\theta}\delta(t)dt = \int_a^b \Re(g(t))dt.$$

In turn, this is

$$\leq \int_a^b |e^{i\theta}\delta(t)|dt = \int_a^b |\delta(t)|dt\,.$$

□

2.2 Change of Parameter and a Fundamental Inequality

Here we deal with the way contour integrals are affected by the various types of parameter changes which can arise.

Let γ be a piecewise smooth curve defined on $[a,b]$ in a domain Ω and $\phi : [\alpha,\beta] \to [a,b]$ be a smooth diffeomorphism taking endpoints to endpoints. We call such a ϕ a *parameter change.* We then get a new (reparameterized) piecewise smooth curve $\gamma \circ \phi : [\alpha,\beta] \to \Omega$ with the same trajectory. It is closed, or simple, if and only if γ is closed, or simple respectively. Observe that ϕ' is never zero. Hence by continuity of ϕ' and connectedness of $[\alpha,\beta]$, we see ϕ' is everywhere positive, or everywhere negative. We then say ϕ is respectively orientation preserving, or orientation reversing. For example, γ_4 is an orientation reversing reparameterization of γ_1.

If γ is a piecewise smooth curve defined on $[a,b]$ in a domain Ω, we can run the curve backwards by means of the change of variable $\phi : [b,a] \to [a,b]$ defined by $\phi(t) = a+b-t$. We call this reparameterized curve $-\gamma$.

Lemma 2.2.1 *The arc length of a piecewise smooth curve is independent of parameterization.*

Proof. $L(\gamma\circ\phi) = \int_\alpha^\beta |(\gamma\circ\phi)'(s)|ds$. But, $(\gamma\circ\phi)'(s) = \gamma'(\phi(s))\phi'(s)$. Hence $|(\gamma\circ\phi)'(s)| = |\gamma'(\phi(s))||\phi'(s)|$. By the change of variable formula of one variable calculus this means

$$L(\gamma \circ \phi) = \int_{\phi(\alpha)}^{\phi(\beta)} |\gamma'(t)| dt = L(\gamma). \qquad \square$$

For γ, a piecewise smooth curve in a domain Ω and $f : \Omega \to \mathbb{C}$ a continuous function, we now define the *contour integral* of f along γ. This will be written $\int_\gamma f$, or sometimes $\int_\gamma f(z)dz$. It is a complex number. It can be defined by taking finite subdivisions of $[a, b]$, forming the analogue of Riemann sums and taking the limit as the mesh of the subdivision tends to zero. If we were dealing with more general contours such as rectifiable ones, this is what we would have to do, but since we will need only piecewise smooth contours and we also want practical formulas for contour integration, we keep it simple. We define

$$\int_\gamma f = \int_a^b f(\gamma(t))\gamma'(t)dt.$$

Clearly, $\int_\gamma f$ is linear in f. Another easily verifiable fact is the following: If $\gamma_1, \ldots \gamma_n$ are piecewise smooth contours in Ω such that for each $j = 1, \ldots, n-1$, $\gamma_j(b_j) = \gamma_{j+1}(a_{j+1})$, then we can string these together to a piecewise smooth contour in Ω which we call $\Sigma\gamma_j$. Moreover, for each continuous function f on Ω we have $\int_{\Sigma\gamma_j} f = \sum_j \int_{\gamma_j} f$. We leave it to the reader to verify these statements.

Contour integration satisfies the following *fundamental inequality*, where $\max_\gamma |f|$ is the maximum value of $|f|$ over the (compact) trajectory of γ.

$$\left| \int_\gamma f \right| \leq \max_\gamma |f| L(\gamma).$$

To see this just observe that $|\int_\gamma f| \leq |\int_a^b f(\gamma(t))\gamma'(t)dt|$. By Lemma 2.1.3, this in turn is

$$\leq \int_a^b |f(\gamma(t))||\gamma'(t)|dt \leq \max_\gamma |f| \int_a^b |\gamma'(t)|dt.$$

But this is just $\max_\gamma |f| L(\gamma)$.

Our next result follows immediately from the fundamental inequality.

Corollary 2.2.2 *Let f_n be a sequence of continuous functions on Ω which converge uniformly on the trajectory of γ to a continuous function f. Then $\int_\gamma f_n \to \int_\gamma f$.*

From this, together with the linearity of contour integration, we immediately get the following:

Corollary 2.2.3 *Let f_n be a sequence of continuous functions on Ω and suppose the series $\sum_{n=0}^{\infty} f_n$ is uniformly convergent on the trajectory of γ to a continuous function f. Then $\sum_{n=0}^{\infty} \int_\gamma f_n = \int_\gamma f$.*

It is reasonable to now ask ourselves what the effect is of a parameter change on a contour integral. The corollary below tells us contour integration is invariant.

Corollary 2.2.4 *Let f be a continuous function, γ be a piecewise smooth curve in a domain Ω and ϕ be a parameter change. Then $\int_\gamma f = \int_{\gamma\cdot\phi} f$. In particular, $\int_{-\gamma} f = -\int_\gamma f$.*

This is because

$$\int_a^b f(\gamma(t))\gamma'(t)dt = \int_\alpha^\beta f(\gamma(\phi(s)))\gamma'(\phi(s))\phi'(s)ds$$
$$= \int_\alpha^\beta f(\gamma \circ \phi(s))(\gamma \circ \phi)'(s)ds\,.$$

We also ask ourselves what the effect is on a contour integral of applying a holomorphic function. Our final result tells us that this also leaves contour integration invariant.

Proposition 2.2.5 *Let $g : \Omega \to \Omega'$ be a holomorphic function, where Ω and Ω' are domains. Suppose $\gamma : [a,b] \to \Omega$ is a piecewise smooth curve in Ω and $g \cdot \gamma$ is the transformed contour in Ω'. Then for all continuous functions f on Ω,*

$$\int_\gamma f(g(z))g'dz = \int_{\phi\circ\gamma} f(w)dw\,.$$

Proof. We know

$$\int_\gamma f(g(z))g'(z)dz = \int_a^b f(g(\gamma(t)))g'(\gamma(t))\gamma'(t)dt$$

and

$$\int_{\phi\circ\gamma} f(w)dw = \int_a^b f(g(\gamma(t)))(g\cdot\gamma)'(t)dt\,.$$

Since by the chain rule, $(g\circ\gamma)'(t) = g'(\gamma(t))\gamma'(t)$, these integrals are equal. □

Exercise 2.1 *Let f and g be holomorphic functions on a domain Ω and γ be a piecewise smooth path in Ω joining a and b. Show that*

$$\int_\gamma fg'dz + \int_\gamma gf'dz = f(b)g(b) - f(a)g(a)\,.$$

2.3 Some important examples of contour integration

Let α and β be points of $\mathbb{C}$ and γ be any smooth curve joining them, then for any integer $m \geq 0$, $\int_\gamma z^m dz = \frac{\beta^m - \alpha^m}{m+1}$. In particular, if α and β are equal, that is, if γ is closed, then $\int_\gamma z^m dz = 0$.

This is because $\int_\gamma z^m dz = \int_a^b z(\gamma(t))^m\gamma'(t)dt$. But, $\frac{d}{dt}z(\gamma(t))^{m+1} = (m+1)z(\gamma(t))^m\gamma'(t)$. Since $\gamma(a) = \alpha$ and $\gamma(b) = \beta$, this result follows from the fundamental theorem of calculus.

Completely analogously to the terminology in calculus, we say a function $f : \Omega \to \mathbb{C}$ has a primitive F on Ω, if $F' = f$ everywhere on Ω. We have just discovered the following important principle:

If the function $f : \Omega \to \mathbb{C}$ has a primitive F on Ω, then for any smooth curve γ in the domain $\int_\gamma f = F(\gamma(b)) - F(\gamma(a))$. The proof, using the fundamental theorem of calculus is identical to what we just did.

We can easily extend this to piecewise smooth curves as follows.

Proposition 2.3.1 *If the continuous function $f : \Omega \to \mathbb{C}$ has a primitive F on Ω, then for any piecewise smooth curve γ in the domain $\int_\gamma f = F(\gamma(b)) - F(\gamma(a))$. Thus, $\int_\gamma f$ depends not on γ, but only on the endpoints of γ. In particular, if γ is closed and f has a primitive, then $\int_\gamma f = 0$.*

Proof. We know the result holds for smooth curves. Let γ be a piecewise smooth curve and observe that if t_j are the break points on $[a, b]$. Then

$$\int_\gamma f = \sum_j \int_{\gamma_j} f = \sum_j F(\gamma(t_{j+1})) - F(\gamma(t_j)).$$

But since in this last expression all the middle terms cancel out, we just get $F(\gamma(b)) - F(\gamma(a))$. □

Because in an open disk any compact set is contained in a closed concentric subdisk and therefore in a domain Ω any compact set is contained in a finite union of closed subdisks, the results of Section 1.7 show that it suffices to assume the power series in the proposition below is merely convergent.

Proposition 2.3.2 *Let $f(z) = \sum_{n=0}^{\infty} a_n(z-a)^n$ be a power series convergent uniformly on compacta on a domain Ω, where the $a_n \in \mathbb{C}$, and $a \in \Omega$. Then for any piecewise smooth closed curve γ in Ω, $\int_\gamma f = 0$.*

Proof. Since γ is a closed curve, using linearity, it follows from the above that $\int_\gamma p(z)dz = 0$ for any polynomial p. Taking p_n to be the nth partial sum and applying Corollary 2.2.3 the proposition follows. □

Another key example of a contour integral calculation is

$$\int_\gamma (z-a)^m dz,$$

where m is a negative integer and γ is a positively oriented circle centered at a of radius $r > 0$. So $\gamma(t) = a + re^{it}$, where $0 \le t \le 2\pi$ and $r > 0$. Hence

$$\int_\gamma (z-a)^m dz = \int_0^{2\pi} (a + re^{it} - a)^m rie^{it} dt.$$

But this last term is

$$r^{m+1} i \int_0^{2\pi} e^{(m+1)it} dt \,.$$

Thus if $m = -1$, we get $i \int_0^{2\pi} dt = 2\pi i$ and if $m \neq -1$, then since $\frac{d}{dt} \frac{e^{i(m+1)t}}{m+1} = ie^{i(m+1)it}$ we get zero. Summarizing, we have

Corollary 2.3.3 *For a positive integer n,*

$$\frac{1}{2\pi i} \int_\gamma \frac{dz}{(z-a)^n}$$

is 0, if $n > 1$, or 1, if $n = 1$.

Of course, if n is negative we already know from the above that the integral is zero.

2.4 The Cauchy theorem in ∗-shaped and simply connected domains

We now prove the converse of Proposition 2.3.1.

Proposition 2.4.1 *Let Ω be a domain and $f : \Omega \to \mathbb{C}$ a continuous function. Suppose for all piecewise smooth curves γ in Ω, $\int_\gamma f$ depends only on the endpoints of γ and not on γ itself. Then f has a primitive on Ω.*

Proof. Let $a \in \Omega$ be fixed. For each $z \in \Omega$ join z to a by a continuous path. Then cover this path as above by a finite number of disks completely contained in Ω (cf. Figure 2.1). Join the centers of successive disks by straight line segments. Each of those segments is in the union of the two disks from which it came. Therefore the whole assemblage lies in Ω. This means a and z can be joined by a piecewise linear arc. In particular, they can be joined by a piecewise smooth arc lying completely in the domain (cf. Figure 2.1). Denote by γ_z the arc joining a to z. Since a and f are fixed, our hypothesis guarantees that $\int_{\gamma_z} f$ depends

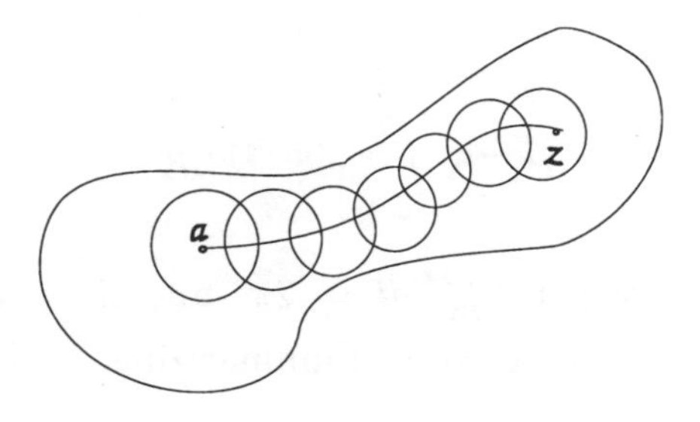

Figure 2.1

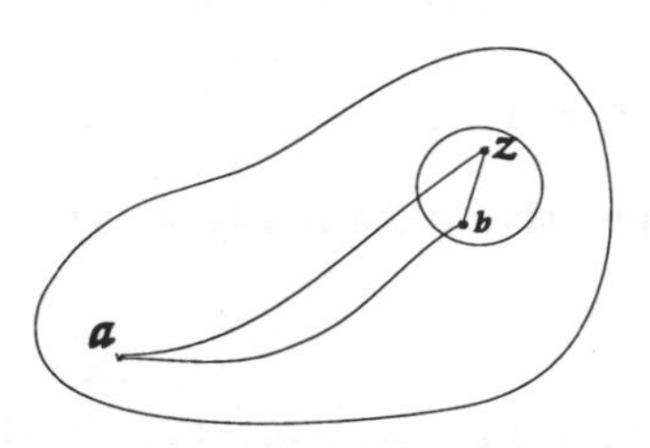

Figure 2.2

only on $z \in \Omega$ and gives a complex valued function on the domain. This will be our primitive, F.

We will show that $F(z) = \int_{\gamma_z} f$ is holomorphic on Ω and $F' = f$, To do this we prove these statements at every point b in the domain.

$$\frac{F(z) - F(b)}{z - b} = \frac{\int_{\gamma_z} f - \int_{\gamma_b} f}{z - b}.$$

Since we are interested in the limit of this for z near b, we may as well assume z starts out life in a small ball in the domain centered at b. Then we join z to b by a straight line segment and join b and z to a as above (cf. Figure 2.2). Since, as we saw, joining piecewise smooth paths at a point where they coincide again gives a piecewise smooth path and since we now have two piecewise smooth paths joining a and b, by the hypothesis of path independence we get $\int_a^b + \int_b^z = \int_a^z$. That is, $\int_a^z - \int_a^b = \int_b^z$. Hence,

$$\frac{F(z) - F(b)}{z - b} = \frac{\int_b^z}{z - b} = \frac{\int_0^1 f(tz + (1-t)b)(z-b)dt}{z - b}$$
$$= \int_0^1 f(tz + (1-t)b)dt\,.$$

So it remains to see that $\lim_{z\to b} \int_0^1 f(tz+(1-t)b)dt = f(b)$. Applying the mean value theorem for integrals to the real and imaginary parts of $\int_0^1 f(tz + (1-t)b)dt$ tells us

$$\int_0^1 f(tz + (1-t)b)dt = \Re f(z_1)(1-0) + i\Im f(z_2)(1-0),$$

where z_1 and z_2 lie on the line segment joining z and b. But since $z \to b$, these points must also approach b. By continuity of f, the limiting value of $\Re f(z_1) + i\Im f(z_2) = f(b)$. Hence $F'(b) = f(b)$ for every $b \in \Omega$. □

Our next result is a precursor of Morera's theorem. Notice that it does not say when either of the two conditions expressed there occurs, just that they are equivalent.

Corollary 2.4.2 *Let f be a continuous function on a domain Ω. Then f has a primitive on the domain if and only if $\int_\gamma f = 0$ for every piecewise smooth closed path in Ω.*

Proof. In view of Propositions 2.3.1 and 2.4.1 it only remains to show that the following conditions are equivalent.

1. For every piecewise smooth closed path in Ω, $\int_\gamma f = 0$.

2. For every piecewise smooth closed path in Ω, $\int_\gamma f$ is path independent.

As we already know the second condition implies the first. Conversely, let α and $\beta \in \Omega$. As above, let γ be a piecewise smooth path in Ω joining them. If δ is another such curve, then $\gamma - \delta$ is a piecewise smooth closed path in Ω. Hence, by assumption, $\int_{\gamma-\delta} f = 0$. But

$$\int_{\gamma-\delta} f = \int_{\gamma+(-\delta)} f = \int_{\gamma} f - \int_{\delta} f.$$

Thus $\int_{\gamma} f = \int_{\delta} f$, proving path independence. □

We can already draw the following conclusion from Corollary 2.4.2.

Corollary 2.4.3 *For any domain Ω and function $f(z) = \sum_{n=0}^{\infty} a_n(z-a)^n$ given by a uniformly convergent power series on the domain, f has a primitive on Ω.*

This follows from Corollary 2.4.2 together with Proposition 2.3.2.

We now come to Cauchy's theorem for a triangle which tells us that under suitable circumstances $\int_{\gamma} f = 0$, if γ is a triangle.

Proposition 2.4.4 *Let f be a holomorphic function on a domain Ω and suppose T is a triangle such that $T \cup \partial T \subseteq \Omega$. Then $\int_{\partial T} f = 0$.*

Proof. By triangle in this proof we mean the closed triangle, i.e. the interior together with the boundary. Divide the triangle into four congruent subtriangles by drawing lines parallel to each side. Orient the original triangle and the four new ones positively (cf. Figure 2.3). Notice that, because of cancellations along the lines that have been drawn, the contour integral of any function over the original triangle is equal

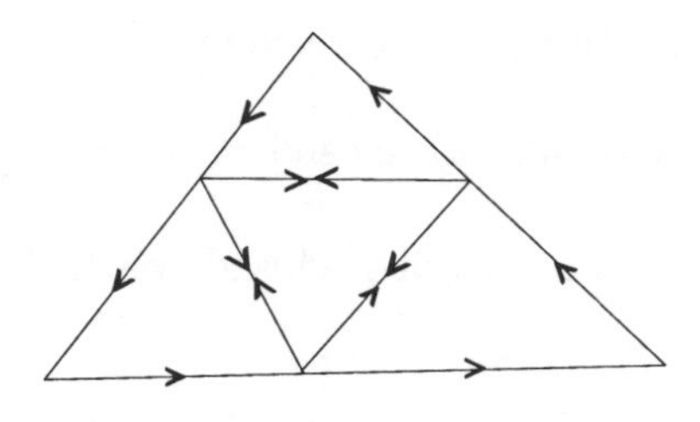

Figure 2.3

to the sum of the contour integrals of the function over the four smaller triangles. Thus

$$\int_{\partial T} f = \sum_{i=1}^{4} \int_{\partial T_i} f.$$

Hence,

$$\left| \int_{\partial T} f \right| \leq \sum_{i=1}^{4} \left| \int_{\partial T_i} f \right| \leq 4 \left| \int_{\partial T^1} f \right|,$$

where T^1 is the subtriangle for which $|\int_{\partial T_i} f|$ is the largest of the four. Continuing in this way we get a nested sequence of triangles $T = T^0 \supseteq T^1 \ldots \supseteq T^n \supseteq \ldots$ such that $|\int_{\partial T^n} f| \leq 4 |\int_{\partial T^{n+1}} f|$ for each n and so

$$\left| \int_{\partial T} f \right| \leq 4^n \left| \int_{\partial T^n} f \right|$$

for all n. By the Bolzano-Weierstrass theorem (completeness) there is a point $a \in \cap_{n=0}^{\infty} T^n$. Since $a \in T \subseteq \Omega$ and f is holomorphic there, we get $f(z) = f(a) + f'(a)(z-a) + \epsilon(z)(z-a)$, $z \in \Omega$ and $\epsilon(z) \to 0$ as $z \to a$.

But then,

$$\begin{aligned} \int_{\partial T^n} f(z)dz &= \int_{\partial T^n} (f(a) + f'(a)(z-a) + \epsilon(z)(z-a))dz \\ &= \int_{\partial T^n} (f(a) + f'(a)(z-a))dz + \int_{\partial T^n} \epsilon(z)(z-a)dz\,. \end{aligned}$$

Since $f(a)+f'(a)(z-a)$ is a polynomial and ∂T^n is closed and piecewise smooth, the first integral is zero (for example, by Corollares 2.4.2 and 2.4.3). As for the second, here we can apply the fundamental inequality and get

$$\begin{aligned} \left| \int_{\partial T^n} \epsilon(z)(z-a)dz \right| &\leq \max |\epsilon(z)|_{\partial T^n} \max |z-a|_{\partial T^n} L(\partial T^n) \\ &\leq \max |\epsilon(z)|_{T^n} \max |z-a|_{T^n} L(\partial T^n)\,. \end{aligned}$$

Since a is in all the T^n and $z \to a$ for any particular given n we may assume $z \in T^n$. Therefore, the above is

$$\leq \epsilon \,\mathrm{diam}(T^n) L(\partial T^n) \leq \epsilon \frac{1}{2^n} \,\mathrm{diam}(T) \frac{1}{2^n} L(\partial T) \,,$$

where here we take n large enough so that on T^n we have $|\epsilon(z)| < \epsilon$. This means

$$\left| \int_{\partial T} f \right| \leq 4^n \left| \int_{\partial T^n} f \right| \leq \epsilon \,\mathrm{diam}(T) L(\partial T) \,.$$

Since $\epsilon > 0$ is arbitrary and $\mathrm{diam}(T)L(\partial T)$ is a constant, $|\int_{\partial T} f| = 0$ and so $\int_{\partial T} f = 0$. □

We now define a $*$-shaped domain. Later we will generalize this notion.

Definition 2.4.5 We say a domain Ω is $*$-*shaped* about a if $a \in \Omega$ and, given any other point $z \in \Omega$, the line segment joining z to a lies completely in Ω. We call a the *center* of the star.

Clearly a convex region is $*$-shaped about each of its points and so these are examples of $*$-shaped regions. A typical non-convex example is provided by removing a closed half line (for example, the negative real axis together with zero) from the plane (cf. Figure 2.4). These regions

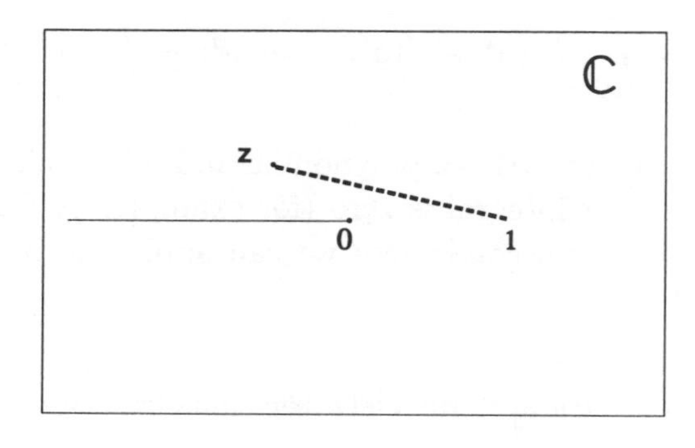

Figure 2.4

are $*$-shaped about each point on the open half line that remains. Also, $*$-shaped regions are contractible by contacting along the lines to the center of the star. In particular, they are simply connected. The student should check these statements.

Our next result is Cauchy's theorem for $*$-shaped domains. Although this itself is an intermediate result, it is sufficient for most situations and gives us something to work with. Another merit it has is that we have essentially already proved it.

Theorem 2.4.6 *If Ω is a $*$-shaped domain and f a holomorphic function on it, then $\int_\gamma f = 0$ for every piecewise smooth closed path γ in Ω.*

Proof. To do this it suffices, by Corollary 2.4.2, to show f has a primitive on the domain. Let a be the center of the star and for $z \in \Omega$, let $F(z) = \int_a^z f$, where the contour of integration is the line segment joining a with z. Thus F is a complex valued function on Ω. Let $b \in \Omega$. We show F is holomorphic at b. Now we argue in a similar way to Proposition 2.4.1.

$$\frac{F(z) - F(b)}{z - b} = \frac{\int_a^z f - \int_a^b f}{z - b}.$$

Since we are interested in the limit of this for z near b we may as well assume z starts out life in a small ball in the domain centered at b. Then we can join z to b by a straight line segment which lies in the ball and hence in the domain. Therefore, we have a triangle and its boundary in the domain. By Cauchy's theorem for a triangle the integral around this is zero. Therefore, $\int_a^z + \int_z^b + \int_b^a = 0$ and since $\int_b^a = -\int_a^b$ we get $\int_a^z - \int_a^b = \int_b^z$. Arguing just as in Proposition 2.4.1, we get $F'(b) = f(b)$. Since this holds for all $b \in \Omega$, F is a primitive for f. □

We can now study the logarithm function in greater detail. Since exp is periodic, it can not be inverted. But, as we saw when we attempted to do so, if $e^w = z$ (of course, here $z \neq 0$) and $w = u + iv$, then $e^w = e^{u+iv} = e^u e^{iv} = z$. Hence $|z| = e^u$ and v is congruent to $\arg z \bmod 2\pi\mathbb{Z}$. So for some integer n, $w = \log(|z|) + i(\arg z + 2\pi n)$. Taking $n = 0$, we get what is called the principal branch of the log. Thus

$$\mathrm{Log}(z) = \log(|z|) + i(\arg z)\,.$$

Now in addition to $z \neq 0$, we want to make the inverse function at least continuous. But going around a circle gives trouble in the formula above. So we exclude this also by taking out a slit through the origin. Now we want a function that agrees with the real log, which is not defined on the negative real axis (or zero). Thus the domain Ω of Log is gotten by excluding the negative part of the real axis together with zero. As we saw above, this is a $*$-shaped domain and in it Log is clearly continuous. If $e^{x+iy} = e^{x'+iy'}$, then as we saw $x = x'$ and $y - y' = 2\pi n$. Therefore $|y - y'| = 2\pi|n|$. If y and $y' \in (-\pi, \pi]$, then $|y - y'| < 2\pi$ and so $|n| = 0$ and $y = y'$. Hence the requirement that $-\pi < \arg z \leq \pi$ gives a domain where Log inverts exp. In particular, since $e^0 = 1$, we know $\mathrm{Log}\, 1 = 0$. Also, the chain rule tells us Log is holomorphic, because exp is, and allows us to calculate its derivative. Since $\exp(\mathrm{Log}(z)) = z$, differentiating and taking into account that the derivative of exp is itself, we get $\exp(\mathrm{Log}(z))\frac{d}{dz}\mathrm{Log}\, z = 1$. So that, just as in the real case,

$$\frac{d}{dz}\mathrm{Log}\, z = \frac{1}{z}\,.$$

This enables us to calculate contour integrals of Log. For example, let w be a point in Ω and γ be any piecewise smooth path in Ω joining 1 with w. Observe that Log has a primitive in Ω. For, using the product rule discussed at the beginning of Section 1.2, we see $\frac{d}{dz}(z\,\mathrm{Log}\, z - z) = \mathrm{Log}\, z$. By Proposition 2.3.1 the integral is independent of γ and only depends on the end points. Evaluating a primitive at those points gives $\int_\gamma \mathrm{Log}(z)dz = w\,\mathrm{Log}\, w - w + 1$.

One more point about the Log function should be made here which shows the importance of the hypothesis of simple connectivity. We ask the reader to check the details: For $z = x + iy \neq 0$, $\omega = \log(|z|)$ is a harmonic function on $\mathbb{C} - (0)$, which is not simply connected. If we restrict ω to our $*$-shaped domain, then it is the real part of a holomorphic function, namely, it is $\Re(\mathrm{Log})$. But ω cannot be the real part of a holomorphic function on $\mathbb{C} - (0)$. For if it were and $f = u + iv$, direct calculation yields $u_x = \frac{x}{x^2+y^2}$ and $u_y = \frac{y}{x^2+y^2}$. Hence, as we saw when we studied the Cauchy-Riemann equations, $v_x = -u_y = \frac{-y}{x^2+y^2}$ and

$$f'(z) = u_x - iu_y = \frac{x - iy}{x^2 + y^2} = \frac{\bar{z}}{|z|^2} = \frac{1}{z}.$$

Since $f'(z) = \frac{1}{z} = \mathrm{Log}'\, z$. This means f and Log differ by a constant on the $*$-shaped domain. But since $f = \mathrm{Log} + c$ extends holomorphically to $\mathbb{C} - (0)$, the same would be true of Log. However, as we saw, Log does not even extend continuously to $\mathbb{C} - (0)$. This is a contradiction.

We conclude this section with Cauchy's theorem for simply connected domains, together with some consequences. Here we generalize Theorem 2.4.6 to the following fundamental result.

A connected domain Ω is called *simply connected* if every piecewise smooth closed curve in it is deformable within Ω to a point. Intuitively, this means the domain has no holes.

Theorem 2.4.7 *If Ω is a simply connected domain and f a holomorphic function on it, then $\int_\gamma f = 0$ for every piecewise smooth closed path γ in Ω.*

Proof. We first consider the case when γ is piecewise linear and forms a *simple* closed polygon (which is not necessarily convex). Give it a positive orientation. Say the polygon has n sides and, of course, n vertices. There is always a vertex from which we can draw $(n-2)$ lines to all the other non-adjacent vertices staying within the polygon and not coinciding with any of its sides, thus creating $n-2$ triangles (cf. Figure 2.5).

Orient these triangles positively. Then, because of cancellation along the lines which we have drawn, the integral around the original polygon is the sum of the integrals around the triangles. Because of simple connectivity, Proposition 2.4.4 applies and tells us the integral around each of the triangles is zero. Hence the integral around our simple, convex closed polygon is also zero. If γ is a closed polygon which is not necessarily simple, then it can be broken up into a finite sum of simple closed polygons. Since the integral around the simple ones is zero (with either positive or negative orientation), the integral around

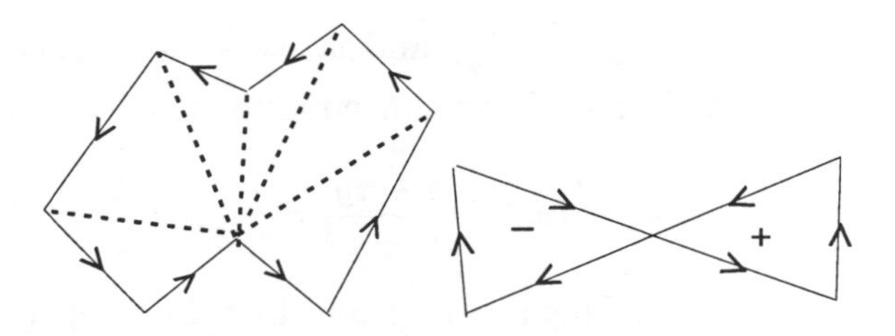

Figure 2.5

γ is also zero. This takes care of the case of a piecewise linear curve γ. Now let $\gamma : [a, b] \to \Omega$ be an arbitrary piecewise smooth closed path and $t_1, \ldots, t_n$ be the finite number of special points in this interval at which the curve may not be smooth. Form subdivisions of $[a, b]$ which include a and b as well as these special points with mesh tending to zero and draw the corresponding chords. This gives a closed polygon, p (which is not necessarily simple), but around which the integral is zero by the above (cf. Figure 2.5). If we can show that for each of these (fixed) finite number of subintervals I, $|\int_{\gamma_I} f - \int_{p_I} f|$ is arbitrarily small, then by adding the results for each subinterval, by the triangle inequality, we would get $|\int_\gamma f - \int_p f| < \epsilon$. Since $\int_p f = 0$ and $\epsilon > 0$ is arbitrary, it would follow that $\int_\gamma f$ is also zero.

Thus we are reduced to studying $|\int_\gamma f - \int_p f|$, where γ is a smooth curve on an interval $I = [c, d]$ and p is a polygonal approximation to it also defined on I with $p(c) = \gamma(c)$ and $p(d) = \gamma(d)$. Now a routine calculation shows that on I,

$$\left|\int_\gamma f - \int_p f\right| \leq \| f \circ \gamma - f \circ p \|_I \, L(p) + \| (f \circ \gamma) \|_I \, |L(\gamma) - L(p)|$$

$$\leq \| f \circ \gamma - f \circ p \|_I \, (|L(\gamma) - L(p)| + L(\gamma))$$

$$+ \| (f \circ \gamma) \|_I \, |L(\gamma) - L(p)| .$$

Let $0 < \epsilon < 1$ and choose subdivisions sufficiently fine so that $\| f \circ p - f \circ \gamma \|_I < \epsilon$ (by uniform continuity) and $|L(\gamma) - L(p)| < \epsilon$

(because γ has finite length). Then

$$\left|\int_\gamma f - \int_p f\right| \leq \epsilon(1 + L(\gamma) + \| (f \circ \gamma) \|_I).$$

Since γ is fixed and therefore $(1 + L(\gamma) + \| (f \circ \gamma) \|_I)$ is constant, this completes the proof. □

Exercise 2.2 *Let f be continuous on Ω and γ_n and γ smooth contours in Ω all defined on $[a,b]$. If γ_n converges to γ in the sense that $\gamma_n \to \gamma$ and $\gamma_n' \to \gamma'$ both uniformly $[a,b]$ (called C^1 convergence), then $\int_{\gamma_n} f \to \int_\gamma f$.*

We remark that, as is evident from the proof, we could just as well have formulated Cauchy's theorem by assuming f to be holomorphic on the interior of Ω and continuous on the boundary. Sometimes it is most conveniently stated in this form.

2.5 Some immediate consequences of Cauchy's theorem for a simply connected domain

Corollary 2.5.1 *Any holomorphic function on a simply connected domain has a primitive.*

This follows immediately from Theorem 2.4.7 together with Corollary 2.4.2.

Corollary 2.5.2 *Let Ω be a simply connected domain and f a holomorphic function on it which never vanishes. Then there is a holomorphic function g on Ω such that $f(z) = e^{g(z)}$, $z \in \Omega$.*

Proof. Since $f \neq 0$ on Ω, $\frac{f'}{f}$ is a holomorphic function on Ω and because the domain is simply connected it has a primitive, F, by Corollary 2.5.1. Thus $F' = \frac{f'}{f}$. Let $h(z) = e^{F(z)}$. Then h is also holomorphic and never vanishes. Moreover, $\frac{d}{dz}(\frac{f}{h}) = \frac{hf' - fh'}{h^2}$. But $h' = e^{F(z)}F' = \frac{hf'}{f}$. Thus $hf' = fh'$ and so $\frac{d}{dz}(\frac{f}{h}) = 0$. It follows from Proposition 1.5.2 that

$\frac{f}{h} = c$, a constant, which is non-zero since both f and h are non-zero. Therefore $c = e^b$ and so $f(z) = e^b e^{F(z)} = e^{b+F(z)}$. Taking $g(z) = b + F(z)$ gives the result. □

In particular for n an integer ≥ 2, taking $e^{\frac{g(z)}{n}}$, we get the following:

Corollary 2.5.3 *Any non-vanishing holomorphic function on a simply connected domain has a holomorphic n^{th} root.*

We have just seen that in the absence of simple connectivity the following result is false.

Corollary 2.5.4 *Let ω be a C^2 harmonic function on a simply connected domain, Ω. Then $\omega = \Re(f)$, where f is holomorphic.*

Proof. Let $f = \omega_x - i\omega_y$. Then f is smooth. Moreover, $(\omega_x)_x = -(\omega_y)_y$ (since ω is harmonic) and $(\omega_x)_y = -(\omega_y)_x$ (since $\omega_{xy} = \omega_{yx}$). Therefore, by the Cauchy-Riemann equations f is holomorphic. Because Ω is simply connected f has a primitive, $F = u + iv$ on Ω by Corollary 2.5.1. Since $F' = f$, by the Cauchy-Riemann equations we get $u_x = \omega_x$ and $u_y = \omega_y$. Now u and ω are differentiable functions and the domain is connected. It follows that they differ by a real constant $u - c = \omega$. Therefore,

$$\Re(F - c) = \Re(u - c + iv) = u - c = \omega\,,$$

where $F - c$ is holomorphic on Ω. □

Our next result generalizes the fact that $\log|z|$ is harmonic on $\mathbb{C}^\times$. It will be useful later.

Corollary 2.5.5 *Let f be a holomorphic function on a domain Ω which is never zero. Then $\log|f(z)|$ is harmonic on Ω.*

Proof. Since being harmonic is a local property, we can restrict ourselves to an arbitrary disk $D \subseteq \Omega$. Now $f \neq 0$ on D and D is simply connected so, by Corollary 2.5.2, $f(z) = e^{g(z)}$, where g is holomorphic on D. Write $g(z) = u(z) + iv(z)$. Then $f(z) = e^{u(z)}e^{iv(z)}$ so that $|f(z)| = e^{u(z)}$ and $\log|f(z)| = u(z)$, which is harmonic as it is the real part of a holomorphic function. □

Corollary 2.5.6 *Let $f : \Omega_1 \to \Omega$ be a holomorphic function, where $f = u + iv$, and $\omega : \Omega \to \mathbb{R}$ be a harmonic function. Then $\omega(u(z), v(z))$, $z \in \Omega_1$, is also harmonic.*

Proof. Let D be a disk in Ω. Then $f^{-1}(D) \cap \Omega_1$ is an open subset of Ω_1 and of course the restriction of f to it is holomorphic. Also ω restricted to D is harmonic. If we can show $\omega(u(z), v(z))$, where $z \in f^{-1}(D) \cap \Omega_1$ is harmonic, we would be done since Ω is a union of such disks. Thus we may assume Ω itself is a disk, that is, we may assume Ω is simply connected. Hence $\omega = \Re(g)$, where g is holomorphic. Therefore, $g \circ f$ is holomorphic on $f^{-1}(D) \cap \Omega_1$. Since $g(f(z)) = \Re(g)(u, v) + i\Im(g)(u, v) = \omega(u, v) + i\Im(g)(u, v)$, we see $\omega(u, v)$ is harmonic. $\square$

Chapter 3

The Main Consequences of Cauchy's theorem

3.1 The Cauchy theorem in multiply connected domains and the pre-residue theorem

We now consider a more complicated situation. Instead of a simply connected domain, Ω, we consider multiply connected domains. We shall content ourselves with the following situation which will be sufficiently general to encompass everything we need. Take a simply connected domain Ω^* and a finite number of mutually non-intersecting, smooth simple closed curves, $\gamma_1, \ldots \gamma_n$, in it. Let Ω be the domain gotten from Ω^* by removing the interiors together with the boundaries of the γ_j (cf. Figure 3.1). What remains is clearly open and connected. The student may wish to prove this by induction on n. For example, if f is holomorphic on Ω^*, except for a finite number of distinct points $p_1 \cdots p_n$, then it is natural to choose mutually non-intersecting small circles, γ_j in this domain centered at the points p_j. Eliminating these circles together with their interiors gives us such a domain.

Now a domain Ω of this general type is clearly not simply connected. In fact, its fundamental group $\pi_1(\Omega)$ is isomorphic with the free group on n generators. Moreover, the boundary of Ω is the boundary of Ω^*, if

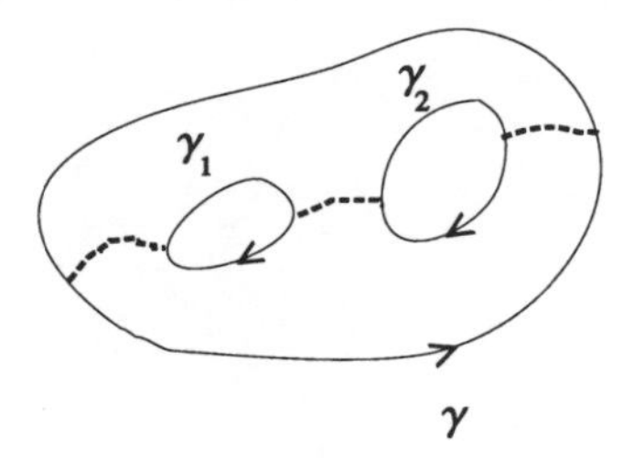

Figure 3.1

any, together with the union of the γ_j. If Ω^* is a bounded domain, then $\partial(\Omega^*)$ is non-empty. Here we shall assume $\partial(\Omega^*)$ is a piecewise smooth simple closed curve γ, which we orient positively. We also orient the γ_j negatively. Thus $\partial(\Omega)$ has $n+1$ components,

$$\partial(\Omega) = \gamma - (\gamma_1 \cup \ldots \cup \gamma_n) .$$

Choose a point on the trajectory of γ and join it to a nearby point on the trajectory of γ_1 by a smooth curve in Ω, as in the diagram. Then, in a similar manner, join γ_j to γ_{j+1}. Finally join γ_n to a point of γ. These curves are called crosscuts. Collectively they divide Ω into two disjoint domains, Ω_1 and Ω_2. Since the boundaries of each of these are piecewise smooth, simple closed curves, Ω_1 and Ω_2 are simply connected by the Jordan Schoenflies theorem mentioned at the beginning of Section 2.1.

Because of cancellation along the crosscuts we see that for a continuous function f,

$$\int_{\partial(\Omega)} f = \int_\gamma f - \sum_{j=1}^{n} \int_{\gamma_j} f = \int_{\partial(\Omega_1)} f + \int_{\partial(\Omega_2)} f .$$

If f is holomorphic on Ω^* and continuous on its boundary, then its restrictions to Ω_1 and Ω_2 have those same properties and so by Theorem 2.4.7 it follows that

$$\int_{\partial(\Omega_1)} f = 0 = \int_{\partial(\Omega_2)} f .$$

Thus we get the Cauchy theorem for multiply connected domains, which generalizes the one for simply connected domains by means of a good organization of the boundary.

Theorem 3.1.1 *If Ω is a multiply connected domain and f a holomorphic function on it, which is continuous on $\partial(\Omega)$, then* $\int_{\partial(\Omega)} f = 0$.

From this follows what we might call the pre-residue theorem in the sense that all the ingredients for the residue theorem are there except for the fact that we have not yet defined residues!

Corollary 3.1.2 *Let Ω be a simply connected bounded domain with boundary γ, and f be continuous on γ and holomorphic on Ω except for a finite number of distinct points $p_1 \cdots p_n$. Choose mutually non-intersecting small circles γ_j in this domain centered at the p_j. Then*

$$\int_\gamma f = \sum_{j=1}^{n} \int_{\gamma_j} f .$$

Finally, we remark that using Corollary 3.1.2 above, together with partial fractions we can calculate a number of contour integrals of rational functions. For example, if γ is any piecewise smooth positively oriented, simple closed curve containing z_1 and z_2 in its interior, then $\int_\gamma \frac{az+b}{(z-z_1)(z-z_2)} = 2\pi i a$. This holds whether or not z_1 and z_2 are distinct. The reader is invited to check both cases.

3.2 The Cauchy integral formula and its consequences

This theorem states the remarkable fact that the boundary values of a holomorphic function on a simply connected domain completely determine the function in the domain. This is quite different from the situation for real functions. Actually, the Cauchy integral formula is a generalization of something we already know (see Section 2.3). Namely, if f is the function 1, then

$$f(z) = \frac{1}{2\pi i}\int_\gamma \frac{1}{\zeta - z}d\zeta . \quad = 1$$

Of course, this fact will play a role in the proof.

We leave it to the reader to formulate a statement and proof of the analogue of the Cauchy integral formula for multiply connected domains along the lines of Theorem 3.1.1, by taking into account those holes which are inside γ.

Theorem 3.2.1 *Let f a holomorphic function on a simply connected domain Ω and γ be a piecewise smooth, positively oriented, simple closed curve in Ω. Then for all z in the interior of γ,*

$$f(z) = \frac{1}{2\pi i}\int_\gamma \frac{f(\zeta)}{\zeta - z}d\zeta .$$

Proof. Here z is fixed and inside γ. For ζ on the trajectory of γ we have, by linearity,

$$\frac{1}{2\pi i}\int_\gamma \frac{f(\zeta)}{\zeta - z}d\zeta = \frac{1}{2\pi i}\int_\gamma \frac{f(z)}{\zeta - z}d\zeta + \frac{1}{2\pi i}\int_\gamma \frac{f(\zeta) - f(z)}{\zeta - z}d\zeta,$$

where $\frac{*}{\zeta - z}$ is a continuous function on the trajectory of γ. Hence, since $\int_\gamma \frac{f(z)}{\zeta - z}d\zeta = \frac{f(z)}{2\pi i}\int_\gamma \frac{1}{\zeta - z}d\zeta$, we see that

$$\frac{1}{2\pi i}\int_\gamma \frac{f(\zeta)}{\zeta - z}d\zeta = \frac{f(z)}{2\pi i}\int_\gamma \frac{1}{\zeta - z}d\zeta + \frac{1}{2\pi i}\int_\gamma \frac{f(\zeta) - f(z)}{\zeta - z}d\zeta.$$

Now since the interior of γ is a simply connected domain, we can use the pre-residue theorem above to replace γ by a small circle γ_r of radius $r > 0$ centered at z and lying wholly inside γ.

$$\frac{1}{2\pi i}\int_\gamma \frac{f(\zeta)}{\zeta - z}d\zeta = \frac{f(z)}{2\pi i}\int_{\gamma_r} \frac{1}{\zeta - z}d\zeta + \frac{1}{2\pi i}\int_{\gamma_r} \frac{f(\zeta) - f(z)}{\zeta - z}d\zeta .$$

But as we know, $\frac{f(z)}{2\pi i}\int_{\gamma_r} \frac{1}{\zeta - z}d\zeta = f(z)$. So it just remains to show the last term is actually zero.

By our fundamental estimate,

$$\left|\frac{1}{2\pi i}\int_{\gamma_r}\frac{f(\zeta)-f(z)}{\zeta-z}d\zeta\right| \le \frac{1}{2\pi}\max_{\zeta\in\Gamma_r}|f(\zeta)-f(z)|\frac{1}{r}2\pi r\,.$$

Thus the integral is $\le \max_{\zeta\in\Gamma_r}|f(\zeta)-f(z)|$, where Γ_r is the trajectory of γ_r. Given $\epsilon>0$, by continuity of f at z, we can choose r small enough so that, if $|\zeta-z|<r$, then $|f(\zeta)-f(z)|<\epsilon$. Since $\epsilon>0$ is arbitrary, this proves

$$\frac{1}{2\pi i}\int_{\gamma_r}\frac{f(\zeta)-f(z)}{\zeta-z}d\zeta = 0\,.$$ □

When we specialize this result to the disk itself we get the so-called Mean Value theorem for holomorphic functions.

Corollary 3.2.2 *Let $D(a,r_0)$ be the disk centered at a of radius $r_0>0$ and f be a holomorphic function on $D(a,r_0)$ and continuous on its boundary. Then for any $r\le r_0$, we have*

$$f(a)=\frac{1}{2\pi}\int_0^{2\pi}f(a+re^{it})dt\,.$$

Proof. By the Cauchy integral formula we get $f(a)=\frac{1}{2\pi i}\int_{\gamma_r}\frac{f(\zeta)}{\zeta-a}d\zeta$. But this integral is just $\frac{1}{2\pi i}\int_0^{2\pi}\frac{f(a+re^{it})}{a+re^{it}-a}ire^{it}dt$. Cancelling out appropriate terms gives the result. □

The following corollary is the mean value theorem for harmonic functions. It actually characterizes harmonic functions, but we shall not interrupt the exposition for this. Observe that this result, being local, does not require simple connectivity.

Corollary 3.2.3 *Let Ω be a domain and u be a harmonic function on it. Then for any $r>0$ for which $D(a,r)\subseteq\Omega$ we have $u(a)=\frac{1}{2\pi}\int_0^{2\pi}u(a+re^{it})dt$.*

This is because u is harmonic on the disk and since a disk is simply connected u is the real part of a holomorphic function f on the disk by Corollary 2.5.4. Using the mean value theorem for holomorphic functions above and taking the real part of both sides, we get Corollary 3.2.3.

Corollary 3.2.4 *A harmonic function ω on a bounded domain Ω takes its minimal and maximal values on the boundary. In particular, if such a function is identically zero on the boundary, then it is identically zero on the domain. Therefore, if two harmonic functions agree on the boundary, they coincide.*

Proof. We prove it for max, min is similar. Suppose this is not the case and ω takes its maximum value at an interior point a. Let D be a disk in Ω centered at a and γ be the positively oriented circle which is the boundary of a slightly smaller disk D, also centerd at a. Then

$$\omega(a) = \frac{1}{2\pi} \int_0^{2\pi} \omega(\gamma(t))dt\,.$$

By the mean value theorem for Riemann integrals $\frac{1}{2\pi}\int_0^{2\pi} \omega(\gamma(t))dt = \omega(\gamma(t_0))$, for some $t_0 \in [0, 2\pi]$. On the other hand, $\gamma(t) \in D$ and so $\omega(a) > \omega(\gamma(t))$ for all $t \in [0, 2\pi]$. This is a contradiction. □

We will now show that if f is holomorphic on a domain Ω then it has all its complex derivatives of all orders. This is also completely different from the situation for real functions. We first need some preliminaries. The lemma below says that $\int_\gamma \psi(\zeta, z)d\zeta$ is a continuous function of z.

Lemma 3.2.5 *Let Ω be a domain, γ be a piecewise smooth curve in it, and $\psi(\zeta, z)$ be a continuous function of two complex variables defined on $\Gamma \times \bar{D}$, where $\zeta \in \Gamma$ and $z \in \bar{D}$, the closed disk, and Γ is the trajectory of γ. Then*

$$\lim_{z\to a} \int_\gamma \psi(\zeta, z)d\zeta = \int_\gamma \psi(\zeta, a)d\zeta\,.$$

Proof. By the fundamental inequality

$$\left|\int_\gamma \psi(\zeta, z)d\zeta - \int_\gamma \psi(\zeta, a)d\zeta\right| \leq L(\gamma) \max_{\zeta\in\Gamma} |\psi(\zeta, z) - \psi(\zeta, a)|\,.$$

Since ψ is uniformly continuous on the compact set $\Gamma \times \bar{D}$, this is less than ϵ for all $\zeta \in \Gamma$ if $|z - a|$ is small enough. □

Proposition 3.2.6 *Let Ω be a domain, γ be a piecewise smooth curve in it, and ϕ a continuous function on Ω. For n a positive integer and $z \in \Omega - \Gamma$ let*

$$\Phi(z) = \frac{1}{2\pi i} \int_\gamma \frac{\phi(\zeta)}{(\zeta - z)^n} d\zeta.$$

Then Φ is holomorphic on $\Omega - \Gamma$ and

$$\Phi'(z) = \frac{n}{2\pi i} \int_\gamma \frac{\phi(\zeta)}{(\zeta - z)^{n+1}} d\zeta.$$

Proof. First note that $\frac{\phi(\zeta)}{(\zeta - z)^n}$ is a continuous function for $\zeta \in \Gamma$ and $z \in \Omega - \Gamma$. As to differentiability

$$\begin{aligned} \frac{\Phi(z) - \Phi(a)}{z - a} &= \frac{1}{2\pi i} \int_\gamma \frac{\phi(\zeta)}{(z - a)} \left(\frac{1}{(\zeta - z)^n} - \frac{1}{(\zeta - a)^n} \right) d\zeta \\ &= \frac{1}{2\pi i} \int_\gamma \frac{\phi(\zeta)}{(z - a)} \frac{((\zeta - a)^n - (\zeta - z)^n)}{(\zeta - z)^n (\zeta - a)^n} d\zeta\,. \end{aligned}$$

Now for a and $b \in \mathbb{C}$ we have $a^n - b^n = (a - b)(a^{n-1} + a^{n-2}b + a^{n-3}b^2 + \ldots ab^{n-2} + b^{n-1})$. So in the integral we get

$$\frac{1}{2\pi i} \int_\gamma \phi(\zeta) \frac{(\zeta - a)^{n-1} + (\zeta - a)^{n-2}(\zeta - z) + \cdots + (\zeta - z)^{n-1}}{(\zeta - z)^n (\zeta - a)^n} d\zeta.$$

Taking the limit as $z \to a$ gives, after cancelling, $\frac{n}{2\pi i} \int_\gamma \frac{\phi(\zeta)}{(\zeta - z)^{n+1}} d\zeta$. □

Of course, in Corollary 3.2.7 below the case $n = 0$ is the Cauchy integral formula itself.

Corollary 3.2.7 *If f is holomorphic on a domain Ω, then it is $C^\infty(\Omega)$. Moreover, if γ is any piecewise smooth, positively oriented, simple closed curve in Ω, then for any z in the interior of γ and $n = 0, 1, \ldots$ we have*

$$f^{(n)}(z) = \frac{n!}{2\pi i} \int_\gamma \frac{f(\zeta)}{(\zeta - z)^{n+1}} d\zeta\,.$$

Proof. Since this is a local question we may assume without loss of generality that the domain is a disk. We will now prove everything under the assumption that the domain Ω is simply connected. By the Cauchy integral formula, $f(z) = \frac{1}{2\pi i}\int_\gamma \frac{f(\zeta)}{\zeta - z}d\zeta$, for all z in the interior of γ. Taking $f = \phi$ and $n = 1$ in the Proposition above we get $f'(z) = \frac{1}{2\pi i}\int_\gamma \frac{f(\zeta)}{(\zeta - z)^2}d\zeta$. Taking $n = 2$ we see that, in turn, $f'(z)$ is itself holomorphic in the domain and $f^{(2)}(z) = \frac{2}{2\pi i}\int_\gamma \frac{f(\zeta)}{(\zeta - z)^3}d\zeta$. Continuing by induction we see f is C^∞ and for all $n \geq 0$, $f^{(n)}(z) = \frac{n!}{2\pi i}\int_\gamma \frac{f(\zeta)}{(\zeta - z)^{n+1}}d\zeta$. □

From this we get the corresponding fact for harmonic functions. Namely, if they have enough derivatives to say they are harmonic then they are actually C^∞ functions. Again, we can look at everything locally.

Corollary 3.2.8 *If ω is a harmonic function on Ω, then it is $C^\infty(\Omega)$.*

We now come to Morera's theorem.

Corollary 3.2.9 *Let f be a continuous function on a domain Ω. If $\int_\gamma f = 0$ for every piecewise smooth closed path in Ω, then f is holomorphic on the domain.*

Proof. Because $\int_\gamma f = 0$ for every piecewise smooth closed path in Ω, Corollary 2.4.2 tells us that f has a primitive, F in the domain. Since F is holomorphic, so is its derivative, namely, f by Corollary 3.2.7. □

From Corollary 3.2.7 we immediately get the corresponding generalization of Corollary 3.2.2.

Corollary 3.2.10 *Let $D(a, r_0)$ be the disk centered at a of radius r_0, where $0 < r_0 \leq \infty$ and f be a holomorphic function on $D(a, r_0)$ and continuous on its boundary. Then for any $r \leq r_0$ and any integer $n \geq 0$, we have*

$$f^{(n)}(a) = \frac{n!}{2\pi i}\int_{\gamma_r} \frac{f(\zeta)}{(\zeta - a)^{n+1}}d\zeta\,,$$

where γ_r is the positively oriented circle of radius r centered at a.

By applying the fundamental inequality, from this Corollary we immediately get the Cauchy inequalities. The only requirement is, naturally that the circle of radius r_0 lie in the domain.

Corollary 3.2.11 *For each $r \leq r_0 \leq \infty$ and integer $n \geq 0$, we have*

$$|f^{(n)}(a)| \leq \frac{n!}{r^n} \parallel f \parallel_{\gamma_r} .$$

Exercise 3.1 1. *Let $f : D(a, r_0) \to \mathbb{C}$ be a holomorphic function on a disk. Show for $0 < r < r_0$ there is a constant M so that for $\zeta \in \bar{D}$*

$$|f^{(n)}(\zeta)| \leq \frac{n! M r_0}{(r_0 - r)^{n+1}} .$$

2. *Now take $r = \frac{r_0}{2}$. Using this, show for a suitable constant K,*

$$\parallel f^{(n)} \parallel_{D(a,r)} \leq K n! \left(\frac{1}{r}\right)^n .$$

3. *Prove the following partial converse to the previous statement: Let f be a real C^∞ function on I (an interval, or a half line, or the whole real axis). Suppose there are positive constants c and K so that, for n sufficiently large, $\parallel f^{(n)} \parallel_I \leq \frac{Kn!}{c^n}$. Then f is real analytic.*

Liouville's theorem, in turn, follows from the first Cauchy inequality.

Corollary 3.2.12 *A bounded entire function must be constant.*

Proof. Since f is entire, we take $r_0 = \infty$. For each $a \in \mathbb{C}$ we have $|f'(a)| \leq \frac{\|f\|_{\mathbb{C}}}{r}$. Letting $r \to \infty$, it follows that $|f'(a)| = 0$. Since this holds for all a, we get $f' \equiv 0$. As we saw earlier this means f must be constant. □

From this follows the Fundamental Theorem of Algebra

Corollary 3.2.13 *Every non-constant polynomial $p(z) = \sum_{j=0}^{n} a_j z^j$ has a zero.*

Of course, using the division algorithm and induction on the degree, it follows that any polynomial factors completely $p(z) = a_n(z - z_1)^{n_1} \dots (z - z_r)^{n_r}$, where $z_1, \dots z_r$ are the distinct roots and $n_1, \dots n_r$ are their multiplicities.

Proof. Here $n \geq 1$ and $a_n \neq 0$. We may clearly assume $a_n = 1$. An easy use of the triangle inequality shows that $\lim_{|z| \to \infty} \frac{|p(z)|}{|z|^n} = 1$. Suppose p has no zero. Then it also follows that $\lim_{|z| \to \infty} \frac{|z|^n}{|p(z)|} = 1$. If $\epsilon > 0$, then for $|z|$ large we get $|\frac{|z|^n}{|p(z)|} - 1| < \epsilon$. In other words, $1 - \epsilon < \frac{|z|^n}{|p(z)|} < 1 + \epsilon$. In particular, letting $|z| = r$, we get for r large that $\frac{1}{|p(z)|} < \frac{1+\epsilon}{r^n}$, so $\frac{1}{|p(z)|}$ is bounded outside of a disk of radius r. On the other hand, $\frac{1}{p(z)}$ is continuous and therefore certainly bounded on the closed disk of radius r. Hence, $\frac{1}{p(z)}$ is bounded and entire and so is constant, by Liouville's theorem. Since $\frac{1}{p(z)} = c$, it follows that $p(z) = \frac{1}{c}$ is also constant, a contradiction. $\square$

We can use the Cauchy inequalities to characterize polynomials among entire functions, by conditions of *polynomial growth.*

Corollary 3.2.14 *An entire function f is a polynomial of degree $\leq n$ if and only if there is a constant $M > 0$ such that for $|z|$ large, $|f(z)| \leq M|z|^n$.*

Proof. Suppose $p(z) = \sum_{j=0}^{n} a_j z^j$ is a polynomial of degree n. Then, since for $|z|$ large enough and $j < n$ we know $|a_j z^j| \leq |a_n| \frac{|z|^n}{n}$, we see that $|\sum_{j=0}^{n-1} a_j z^j| \leq |a_n||z|^n$. Hence for $|z|$ large enough $|\sum_{j=0}^{n} a_j z^j| \leq 2|a_n||z|^n$.

Conversely, suppose for $|z|$ large $|f(z)| \leq M|z|^n$ (here n is fixed and z varies). Since f is entire, by Cauchy's inequalities, for $a \in \mathbb{C}$ we see that

$$|f^{(n+1)}(a)| \leq \frac{(n+1)!}{r^{n+1}} \parallel f \parallel_{\gamma_r} \leq \frac{(n+1)!}{r^{n+1}} M \max_{\gamma_r}(|z|)^n .$$

Hence $|f^{(n+1)}(a)| \leq (n+1)! M (\frac{\max_{\gamma_r} |z|}{r})^n \frac{1}{r}$. If we can show that $(\frac{\max_{\gamma_r} |z|}{r})^n$ is bounded, then by taking $r \to \infty$ we would get $f^{(n+1)} = 0$. Hence f would be a polynomial of degree $\leq n$ and we would be done.

But if $r = |z-a|$, by the triangle inequality we see $|a| + r \geq |z|$. Hence $(\frac{|a|}{r}+1)^n \geq (\frac{|z|}{r})^n$. Since a is fixed, as $r \to \infty$ $(\frac{|a|}{r}+1)^n \to 1$. So $(\frac{\max_{\gamma_r}|z|}{r})^n$ is bounded. □

3.3 Analyticity, Taylor's theorem and the identity theorem

Theorem 3.3.1 *Let Ω be a domain and f be a holomorphic function on it. Let $a \in \Omega$ be fixed. Choose the* **largest** *open disk $D(a,r)$ centered at a and lying in Ω (here $0 < r \leq \infty$). Then on $D(a,r)$ we have $f(z) = \sum_{n=0}^{\infty} a_n (z-a)^n$, where $a_n = \frac{f^{(n)}(a)}{n!}$.*

This is called the Taylor series of f at a and the a_n are called the Taylor coefficients. They are uniquely determined by f (and a).

Proof. Let γ be the circle of radius r centered at a, where we take $0 < r < \infty$. For ζ on the trajectory of γ and $z \in D(a,r)$, $\zeta - z \neq 0$ so we can form the continuous function $\frac{1}{\zeta - z}$ (cf. Figure 3.2). In particular, $\frac{1}{\zeta - a}$ is continuous. An easy calculation shows

$$\frac{1}{\zeta - z} = \frac{1}{\zeta - a} \cdot \frac{1}{1 - \frac{z-a}{\zeta - a}}.$$

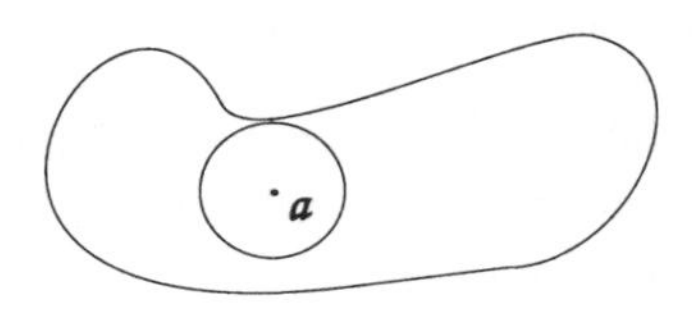

Figure 3.2

Since $|z-a| < |\zeta - a|$, we get $|\frac{z-a}{\zeta-a}| < 1$. Hence by Proposition 1.7.1 this series converges, and so

$$\frac{1}{1-\frac{z-a}{\zeta-a}} = \sum_{n=0}^{\infty}\left(\frac{z-a}{\zeta-a}\right)^n .$$

It follows that

$$\frac{f(\zeta)}{\zeta - z} = \frac{f(\zeta)}{\zeta - a}\sum_{n=0}^{\infty}\left(\frac{z-a}{\zeta-a}\right)^n = \sum_{n=0}^{\infty}\frac{f(\zeta)(z-a)^n}{(\zeta-a)^{n+1}} .$$

We now use the uniform convergence on compacta of the geometric series and therefore the one gotten by multiplying by $f(\zeta)$ to integrate by $\frac{1}{2\pi i}\int_\gamma$ term by term. Applying the Cauchy integral formula gives

$$f(z) = \sum_{n=0}^{\infty}(z-a)^n\frac{1}{2\pi i}\int_\gamma \frac{f(\zeta)}{(\zeta-a)^{n+1}}d\zeta .$$

This gives our result where $a_n = \frac{1}{2\pi i}\int_\gamma \frac{f(\zeta)}{(\zeta-a)^{n+1}}d\zeta$. But, by Corollary 3.2.7, this is nothing more than $\frac{f^{(n)}(a)}{n!}$. □

Now the coefficients are unique by the very fact that we have a formula for them in terms of f and a. Another way to see the uniqueness is to use term by term differentiation of the power series n times and then to evaluate at a. So if $f(z) = \sum_{n=0}^{\infty} b_n(z-a)^n$, then $f'(z) = \sum_{n=1}^{\infty} nb_n(z-a)^{n-1}$. Therefore $b_1 = f'(a) = a_1$, etc. Later in this section we will explain why term by term differentiation of a power series gives the derivative of the function.

Corollary 3.3.2 *A holomorphic function is analytic.*

As explained earlier, a function with a convergent Taylor series at each point is an analytic function. Of course, taking different base points may give different power series.

Corollary 3.3.3 *An entire function is represented by a power series of infinite radius about each point.*

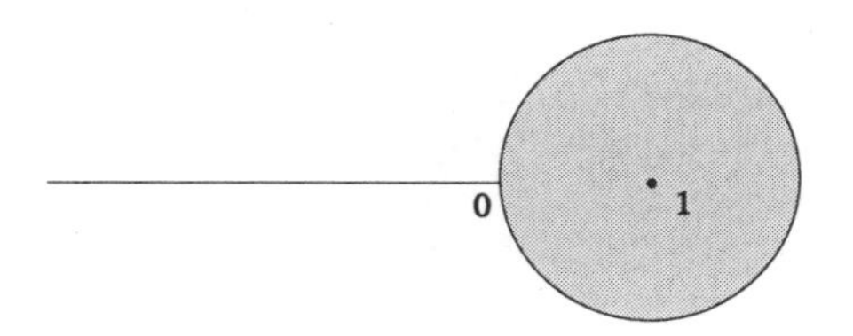

Figure 3.3

We now briefly return to the Log function defined on the $*$-shaped domain, Ω, gotten by removing the negative y-axis together with 0 from $\mathbb{C}$. Recall that $\mathrm{Log}(1) = 0$ and for $z \in \Omega$, $\mathrm{Log}'(z) = \frac{1}{z}$. A direct calculation shows that, for all $n \geq 1$, $\frac{\mathrm{Log}^{(n)}(z)}{n!} = \frac{(-1)^{n-1}}{n} z^{-n}$. Hence, in the largest open disk centered at 1 and contained in Ω, we get the Taylor expansion (cf. Figure 3.3). Thus for $|z-1| < 1$ we get

$$\mathrm{Log}\, z = \sum_{n=1}^{\infty} \frac{(-1)^{n-1}}{n} (z-1)^n .$$

(This is sometimes written $\mathrm{Log}(1+z) = \sum_{n=1}^{\infty} \frac{(-1)^{n-1}}{n} z^n$, where $|z| < 1$.)

Now the Taylor series above is only valid in $D(1,1)$; but the conditions $\mathrm{Log}(1) = 0$ and for $z \in \Omega$, $\mathrm{Log}'(z) = \frac{1}{z}$ uniquely determine Log in all of Ω. For if F also satisfies these conditions, then $F'(z) = \frac{1}{z} = \mathrm{Log}'(z)$. By Corollary 1.5.3, $F - \mathrm{Log} = c$, a constant, and since $F(1) = 0 = \mathrm{Log}(1)$, $c = 0$ and therefore $F = \mathrm{Log}$.

Exercise 3.2 *Show if $|z| < 1$, then $\Re(\frac{1}{1-z}) > 0$. Calculate the power series for $\mathrm{Log}(\frac{1}{1-z})$ about $z = 0$ in the unit disk.*

We now turn to the identity theorem.

Theorem 3.3.4 *Let Ω be a domain and f be a holomorphic function on it. If there is a sequence z_n of distinct points in Ω converging to a point $a \in \Omega$, and $f(z_n) = 0$ for all n, then $f \equiv 0$.*

This means that if we have two holomorphic functions f and g and $f(z_n) = g(z_n)$ for all n, then $f \equiv g$ (cf. Figure 3.4).

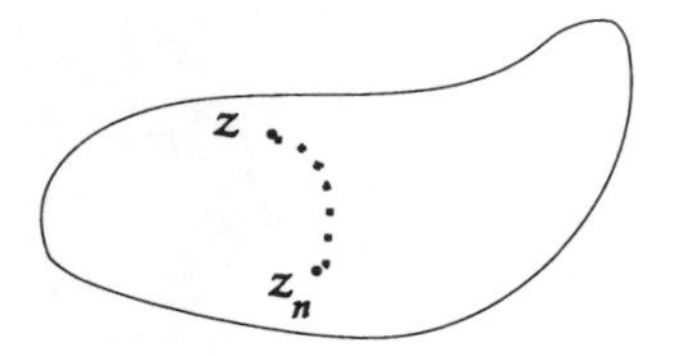

Figure 3.4

Proof. We first consider the case where $\Omega = D(a, r)$. Then by Theorem 3.3.1, $f(z) = \sum_{n=0}^{\infty} a_n(z-a)^n$ for all $z \in D(a, r)$. Since f is continuous and $f(z_n) = 0$ for all n, $f(a) = 0$. Therefore $a_0 = 0$. Assume inductively that $a_0, a_1, \ldots, a_j = 0$. Then

$$f(z) = \sum_{n=j+1}^{\infty} a_n(z-a)^n = (z-a)^{j+1} g(z),$$

where $g(z) = a_{j+1} + a_{j+2}(z-a) + \cdots$. Now $f(z_n) = (z_n - a)^{j+1} g(z_n) = 0$. Since each of the z_n (except perhaps one of them) is different from a, the first factor can not be zero. Therefore $g(z_n) = 0$ for all n. Applying the same reasoning to the analytic function g which we applied to f tells us that $a_{j+1} = 0$. By induction, all $a_n = 0$ and, therefore, $f \equiv 0$ on $D(a, r)$. This proves the result when Ω is a disk.

For the general case. let $D(a, r)$ be a disk in Ω centered at a. By the paragraph above $f \equiv 0$ on $D(a, r)$. Suppose $z \in \Omega - D(a, r)$. Join a and z by a piecewise smooth path γ within Ω, *which is parameterized by arc length.* The boundary of $D(a, r)$ must meet γ (in perhaps several points). Choose one such point, say z_1. Choose a disk $D(z_1, r_1)$ in Ω centered at z_1. Then $D(a, r)$ meets $D(z_1, r_1)$ in a lunette shaped region on which $f \equiv 0$ which contains z_1. Since there is a sequence of distinct points of the lunette shaped region converging to z_1 on which f is zero we get, by the above, $f \equiv 0$ on $D(z_1, r_1)$. We will show that continuing in this manner we must reach z after a finite number of steps. Hence $f(z)$ would be zero and this would complete the proof (cf. Figure 3.5).

Since γ is parameterized by arc length, going further in t is the same as going further along the curve. Suppose we can not reach z by this

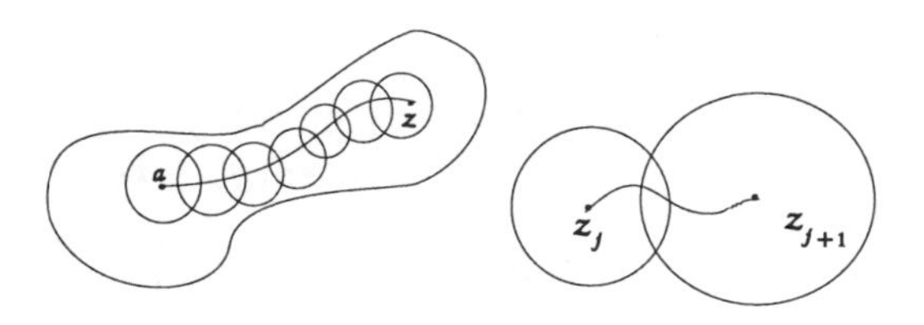

Figure 3.5

process. This means we are approaching a limiting parameter value $\bar{t}$ where $\gamma(\bar{t}) \neq z$. But then by our construction we would get a sequence of distinct points $w_n = \gamma(t_n)$, where $t_n \uparrow \bar{t}$, at which $f(w_n) = 0$. Hence, as above, $f(\gamma(\bar{t})) = 0$. However, our construction tells us we can then go beyond $\gamma(\bar{t})$. This is a contradiction and completes the proof. □

We now get two useful consequences of the identity theorem.

Corollary 3.3.5 *If f is holomorphic and not identically zero, then its zeros are isolated. If the domain is bounded, then its zeros are finite in number.*

Our next corollary explains why, for example, $\sin^2 z + \cos^2 z = 1$ and $\cosh^2 z - \sinh^2 z = 1$. Using this result one variable at a time tells us that because $e^x e^y = e^{x+y}$, it follows that $e^z e^w = e^{z+w}$.

Corollary 3.3.6 *Let $p(w_1, \ldots, w_n)$ be a polynomial in n complex variables and let $w_j = f_j(z)$ be n holomorphic functions defined on the same domain Ω, where $\Omega \supset [a, b]$, a real interval. If $p(f_1(x), \ldots, f_n(x))$ is identically zero for all $x \in [a, b]$, then $p(f_1(z), \ldots, f_n(z))$ is identically zero on Ω.*

We make one additional observation about the identity theorem. That is, it also holds for real analytic functions, f. This is because, as was mentioned earlier, such functions extend in a natural way to the corresponding complex domain (see last paragraph of Section 1.3). Hence, if such a function is zero on a subset of the domain of f in $\mathbb{R}$ with a limit point, this is also true of the holomorphic extension and so the identity theorem forces the extension to be identically zero. Hence f is also zero.

Exercise 3.3 1. *For z and w complex and x and y real, prove using the identity theorem that* $\sin(z+w) = \sin(z)\cos(w) + \cos(z)\sin(w)$, *and therefore,* $\sin(x+iy) = \sin(x)\cosh(y) + i\cos(x)\sinh(y)$.

2. *Let Ω be a domain and f and g holomorphic functions on it. Show if there is a point $a \in \Omega$ such that $f^{(n)}(a) = g^{(n)}(a)$ for all integers $n \geq 0$, then $f \equiv g$.*

3. *Let Ω be a domain and u be a non-negative harmonic function on it. Show if u has a zero, then it is identically zero.*

Now we consider limits of holomorphic functions on a simply connected domain in the topology of uniform convergence on compacta.

Corollary 3.3.7 *If f_n is a sequence of holomorphic functions on a domain Ω and $f_n \to f$ uniformly on compacta of Ω, then f is holomorphic.*

Proof. Since f is the uniform on compacta limit of continuous functions f, is also continuous. Therefore, we can perform contour integration on it. As usual, since being holomorphic is a local property, we may assume our domain is simply connected. Let γ be any piecewise smooth closed curve in Ω. Since the trajectory of γ is compact and so $f_n \to f$ uniformly on it, the fundamental inequality tells us $\int_\gamma f_n \to \int_\gamma f$. By Theorem 2.4.7 we get $\int_\gamma f_n = 0$, for all n. Hence $\int_\gamma f = 0$. By Morera's theorem (Corollary 3.2.9), f is holomorphic. □

Corollary 3.3.8 *For a domain Ω the notions of a function on Ω being holomorphic, complex C^∞ and analytic are all equivalent.*

Proof. A holomorphic function is complex C^∞. This is the content of Corollary 3.2.7. Therefore these two notions are equivalent. By Theorem 3.3.1 any holomorphic function is analytic. We now prove the converse. Since this is a local question, we may assume we are on a disk. But, by the results of Section 1.7, a power series converges uniformly on compacta and polynomials are holomorphic. Hence this follows immediately from Corollary 3.3.7. □

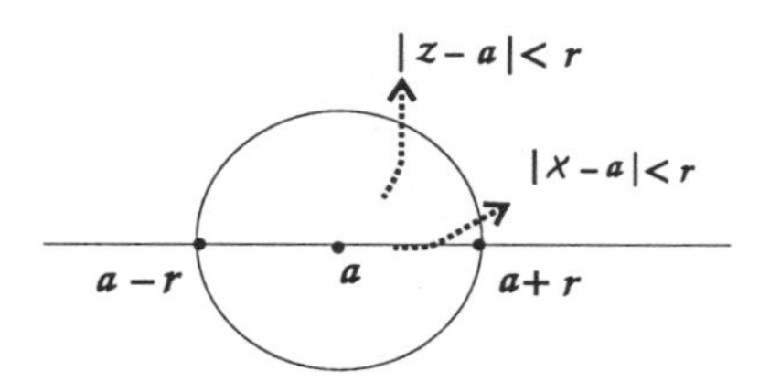

Figure 3.6

Corollary 3.3.9 *The composition of analytic functions is analytic. The composition of real analytic functions is real analytic.*

The first statement follows from Corollary 3.3.8 since the composition of holomorphic functions is holomorphic. The second follows from the first because, as explained earlier, real analytic functions extend to complex analytic functions (cf. Fig. 3.6).

Corollary 3.3.10 *Let f_n be a sequence of holomorphic functions on a domain Ω which converge to f uniformly on compacta of Ω. Then $f_n^{(k)}$ converges uniformly on compacta to $f^{(k)}$.*

Proof. By Corollary 3.3.7, f is holomorphic and therefore by Corollary 3.3.8, it has derivatives of all orders. By induction we need only prove the result for $k = 1$. As we know it suffices to assume $\Omega = D(a, r_0)$ where $r_0 > 0$ and K is a closed subdisk centered at a. For $z \in K$, $f_n'(z) = \frac{1}{2\pi i}\int_\gamma \frac{f_n(\zeta)}{(\zeta - z)^2}d\zeta$ and $f'(z) = \frac{1}{2\pi i}\int_\gamma \frac{f(\zeta)}{(\zeta - z)^2}d\zeta$, where γ is the boundary of K. Since there is some $d > 0$ such that $|\zeta - z| \geq d$ for all ζ on the boundary and $z \in K$, the fundamental inequality tells us that

$$|f_n'(z) - f'(z)| \leq \frac{1}{2\pi} 2\pi r_0 \frac{\| f - f_n \|_K}{d^2},$$

for $z \in K$. But $\| f - f_n \|_K \to 0$. Hence f_n' converges to f' uniformly on K. □

We conclude this section with a preliminary version of Montel's theorem. We will get the final version in Section 3.6.

Definition 3.3.11 Let $\mathcal{F}$ be a family of holomorphic functions on a domain Ω. Suppose for each compact subset $K \subseteq \Omega$ there is a constant $c(K) > 0$ such that $\| f \|_K \leq c(K)$ for all $f \in \mathcal{F}$. We then say the family $\mathcal{F}$ is *locally bounded.*

Let $\epsilon > 0$ and suppose for each point $a \in \Omega$ there is a $\delta(a, \epsilon)$ for which the disk $D_\delta(a)$ in Ω has the following property: If $z \in D_\delta(a)$, then $|f(z) - f(a)| < \epsilon$ for all $f \in \mathcal{F}$. We then say $\mathcal{F}$ is *equicontinuous* at a. If $\mathcal{F}$ is equicontinuous at each point of Ω, we say $\mathcal{F}$ is equicontinuous on Ω.

Corollary 3.3.12 *Let $\mathcal{F}$ be a family of holomorphic functions on a domain Ω which is locally bounded. Then $\mathcal{F}$ is equicontinuous on Ω.*

Proof. Let $a \in \Omega$. We show $\mathcal{F}$ is equicontinuous at a. Let γ be a small positively oriented circle in Ω centered at a of radius $r > 0$. For $z \in D(a, r)$, by Cauchy's integral formula,

$$f(z) = \frac{1}{2\pi i} \int_\gamma \frac{f(\zeta)}{\zeta - z} d\zeta ,$$

for each $f \in \mathcal{F}$. Hence $f(z) - f(a) = \frac{1}{2\pi i} \int_\gamma \frac{f(\zeta)(z-a)}{(\zeta - z)(\zeta - a)} d\zeta$. Therefore, by the fundamental inequality,

$$|f(z) - f(a)| \leq |z - a| \frac{c(\overline{D(a,r)})}{r} ,$$

for all $f \in \mathcal{F}$. Since $\frac{c(\overline{D(a,r)})}{r}$ is a constant, $\mathcal{F}$ is equicontinuous at a. □

3.4 The area formula and some consequences

In this section we will derive a formula for the area of the image of a disk under a 1:1 holomorphic function in terms of the Taylor coefficients of the function.

Lemma 3.4.1 *Let $D(a, r_0)$ be the disk centered at a of radius $0 < r_0 \leq \infty$ and*

$$f(z) = \sum_{n=0}^{\infty} a_n (z - a)^n$$

be an analytic function defined on all of $D(a, r_0)$. *Then for* $0 < r < r_0$

$$\frac{1}{2\pi}\int_0^{2\pi} |f(a+re^{it})|^2 dt = \sum_{n=0}^{\infty} |a_n|^2 r^{2n} .$$

Proof. Let $z(t) = a + re^{it}$, for $0 \le t \le 2\pi$. Then $f(z) = \sum_{n=0}^{\infty} a_n r^n e^{int}$. Therefore,

$$\overline{f(z)} = \sum_{m=0}^{\infty} \bar{a}_m r^m e^{-imt}$$

and so

$$|f(a+re^{it})|^2 = \sum_{n,m=0}^{\infty} a_n \bar{a}_m r^{n+m} e^{i(n-m)t} .$$

Making use of the absolute and uniform convergence of this last series for $r < r_0$ to integrate term by term, together with the orthogonality relations, $\frac{1}{2\pi}\int_0^{2\pi} e^{i(n-m)t} dt = \delta_{n,m}$, we get our conclusion. □

We can now use this to calculate the area of $f(D(a,r))$ when f is 1:1.

Theorem 3.4.2 *Let* $D(a, r_0)$ *be the disk centered at* a *of radius* $r_0 > 0$ *and*

$$f(z) = \sum_{n=0}^{\infty} a_n (z-a)^n$$

be a 1:1 *analytic function defined on* $D(a, r_0)$. *Then for* $0 < r < r_0$ *we have*

$$\text{Area}(f(D(a,r))) = \pi \sum_{n=1}^{\infty} n |a_n|^2 r^{2n} .$$

Proof. We first recall the change of variable formula which is valid for any smooth bijective change of variable, f, defined on a domain D. For all C^∞ functions ϕ with compact support on $f(D)$ we have

$$\int_{f(D)} \phi(w) d\mu(w) = \int_D \phi(f(z)) |\det J_f(z)| d\mu(z) ,$$

where $d\mu(w)$ and $d\mu(z)$ are, respectively, Lebesgue measure on $f(D)$ and D.

Now we take D to be the disk mentioned above and f holomorphic. Then as we saw, $\det J_f(z) = |f'(z)|^2$ by Corollary 1.4.2. Moreover, passing to polar coordinates on D, we have $dxdy = rdrd\theta$. Taking $\phi = 1$ gives

$$\text{Area}(f(D(a,r))) = \int_0^{2\pi} \int_0^r |f'(z)|^2 r dr d\theta \,.$$

Now since $f'(z) = \sum_{n=1}^{\infty} na_n(z-a)^{n-1}$, this means

$$\text{Area}(f(D(a,r))) = \int_0^r rdr \int_0^{2\pi} |f'(z)|^2 d\theta \,.$$

Taking into account Lemma 3.4.1 applied to the holomorphic function f', we get

$$\text{Area}(f(D(a,r))) = \int_0^r rdr 2\pi \sum_{n=1}^{\infty} n^2 |a_n|^2 r^{2(n-1)} = \pi \sum_{n=1}^{\infty} n|a_n|^2 r^{2n} \,. \quad \square$$

Since the area of D itself is πr^2 it is useful to make a comparison of the areas. Under the same hypothesis as Theorem 3.4.2 we have the following

Corollary 3.4.3

$$\frac{\text{Area}(f(D(a,r)))}{\text{Area}(D(a,r))} = \sum_{n=1}^{\infty} n|a_n|^2 r^{2n-2} \,.$$

These considerations give us another way to look at polynomials. Notice that although Theorem 3.4.2 requires that f be 1:1, Lemma 3.4.1 does not.

Corollary 3.4.4 *Let f be entire so for all $z \in \mathbb{C}$ we have $f(z) = \sum_{n=0}^{\infty} a_n z^n$. Suppose f has polynomial growth of order k, that is, $|f(z)| \leq M|z|^k$. Then for each integer n and all $r > 0$,*

$$|a_n| \leq \frac{M}{r^{n-k}} \,.$$

In particular, if $n > k$, taking $r \to \infty$ we get $a_n = 0$.

Proof. By the lemma for any n,

$$|a_n|^2 r^{2n} \leq \sum_{n=0}^{\infty} |a_n|^2 r^{2n} = \frac{1}{2\pi} \int_0^{2\pi} |f(a + re^{it})|^2 dt\,.$$

This, in turn,

$$\leq \frac{1}{2\pi} M^2 \int_0^{2\pi} r^{2k} dt = M^2 r^{2k}$$

by the growth estimate. □

We get some additional corollaries from Theorem 3.4.2. The first we already know from the fact that f is an open map (because f is 1:1, its derivative is never zero).

Corollary 3.4.5 *If* $\text{Area}(f(D)) = 0$, *then* f *is constant.*

This is because $n|a_n|^2 = 0$ for all $n \geq 1$. Therefore $a_n = 0$ for all $n \geq 1$ and so $f(z) = a_0$.

Corollary 3.4.6 *If* $\frac{\text{Area}(f(D))}{\text{Area}(D)}$ *is constant for all* r *(or for all small* r*), then* f *is a linear function and* $|f'(a)| = 1$. *In particular, if* f *preserves area then* f *is in the Euclidean group of the plane.*

Proof. Let $c > 0$ be the constant. Then by Corollary 3.4.3 we get $c - |a_1|^2 = \sum_{n=2}^{\infty} n|a_n|^2 r^{2n-2}$. By the identity theorem $n|a_n|^2 = 0$ and so $a_n = 0$ for all $n \geq 2$. Therefore f is a linear function and $c = |a_1|^2$. If f preserves area, then $c = 1$. Hence, so is $|a_1|$. Thus translating everything to zero, we see f is a rotation. This means f is in the Euclidean group. □

Our final corollary will be needed in the next section. We denote by $\|\|_{2,D}$ the L^2 norm on D with respect to Lebesgue measure. Since here we rely only on the lemma, f need not be 1:1.

Corollary 3.4.7 *For* f *holomorphic on a disk* D *centered at* a *we have* $|f(a)|\sqrt{\text{Area}(D)} \leq \| f \|_{2,D}$.

Proof. Taking the equation which is the conclusion of Lemma 3.4.1, multiplying by r and integrating dr gives

$$\frac{1}{2\pi}\int\int_D |f|^2 d\mu = \sum_{n=0}^{\infty} |a_n|^2 \frac{r^{2n+2}}{2n+2}.$$

Since all terms of the series are positive we get an estimate by taking only the zero term. Then taking square roots yields the desired conclusion. □

Exercise 3.4 *Find a counterexample to the statement of Theorem* 3.4.2 *in case f is not* 1:1.

3.5 Application to spaces of square integrable holomorphic functions

Let $(D, d\mu)$ denote the unit disk with Lebesgue measure and $H^2(D)$ be the holomorphic functions on D which are square integrable with respect to μ. Although in real analysis it is usually necessary to complete a pre-Hilbert space to get a Hilbert space, here we shall see things are different. Of course, we still have to look at classes of functions identified almost everywhere (a.e.) and not merely functions.

For the remainder of this book, given a domain Ω, we let $C(\Omega)$ stand for the space of complex valued continuous functions on Ω. As we shall see in the next section, $C(\Omega)$ is a complete metric space in the topology of uniform convergence on compacta and we will use that fact here.

Theorem 3.5.1 *$H^2(D)$ is actually a* **closed** *subspace of the Hilbert space, $L^2(D, d\mu)$, and therefore is itself a Hilbert space.*

In this connection our first result follows from calculations we made in the area theorem.

Proposition 3.5.2 *For each compact subset K of D there exists a positive constant c_K such that $\| f \|_K \leq c_K \| f \|_2$ for every holomorphic function f on D.*

Proof. Let f be a holomorphic function on D. For each $k \in K$ choose an $r(k) > 0$ so that on the subdisk $D(k, r(k))$ of D the Taylor series for f at k namely $\sum_{n=0}^{\infty} a_n(k)(z-k)^n$ converges absolutely. Since K is compact its points are uniformly bounded away from the boundary of D and hence $\{r(k) : k \in K\}$ is bounded away from zero. It follows that there is an $r_K > 0$ such that $r_K \leq r(k)$ for all $k \in K$. By Corollary 3.4.7 we see that for each fixed $k \in K$,

$$|f(k)|^2 \text{Area}(D(k, r(k))) \leq \| f \|^2_{2, D(k, r(k))} .$$

Therefore,

$$\int_D |f(z)|^2 d\mu \geq \int_{D(k, r_K)} |f(z)|^2 d\mu \geq |f(k)|^2 \mu(D(k, r_K)) .$$

Taking supremum over K gives the required inequality, where $c_K = \frac{1}{\sqrt{\mu(D(k, r_K))}}$, which is independent of $k \in K$ since this is just $\frac{1}{\sqrt{\pi} r_K}$. □

Turning to the proof of Theorem 3.5.1, let f_n be a sequence of holomorphic functions on D converging in L^2 to a function f on D. $\| f_n - f \|_2 \to 0$, where $f \in L^2(D, d\mu)$. If K is a compact subset of D, then by Proposition 3.5.2 there is a constant c_K such that for all n and m,

$$\| f_n - f_m \|_K \leq c_K \| f_n - f_m \|_2 . \tag{3.1}$$

Since f_n is a Cauchy sequence in L^2 and $C(D)$ is complete in the topology of uniform convergence on compacta, it follows that there is a function g on D such that f_n converges uniformly on compacta to g. Hence, by Corollary 3.3.7, we know g is holomorphic. Our task is to show that $g \in L^2(D)$ and $f = g$ a.e. on D. By Eq. (3.1) we see that

$$\| f_n - g \|_K \leq c_K \| f_n - f \|_2 . \tag{3.2}$$

Now f_n converges to f in L^2. It follows that $\| f_n \|_2 \to \| f \|_2$. Hence, if n is large enough, $\| f_n - f \|_2 \leq 1$ and $\| f_n \|_2 \leq \| f \|_2 + 1$.

Let K be a fixed compact subset of D. For n sufficiently large

$$\left(\int_K |g(z)|^2 d\mu\right)^{\frac{1}{2}} \leq \left(\int_K |f_n(z) - g(z)|^2 d\mu\right)^{\frac{1}{2}} + \left(\int_K |f_n(z)|^2 d\mu\right)^{\frac{1}{2}}.$$

Since by (3.2) f_n converges to g uniformly on compacta, $|f_n(z) - g(z)|^2 \leq 1$ on K so that $(\int_K |f_n(z) - g(z)|^2 d\mu)^{\frac{1}{2}} \leq \mu(D)^{\frac{1}{2}}$. Hence, $(\int_K |g(z)|^2 d\mu)^{\frac{1}{2}} \leq \mu(D)^{\frac{1}{2}} + \| f \|_2 + 1$. Thus, for each compact $K \subset D$, $\int_K |g(z)|^2 d\mu \leq c$, where c is a constant independent of K. Choose an increasing sequence K_n of compact disks centered at zero and filling out D. Since for each K_n, $\int_{K_n} |g(z)|^2 d\mu \leq c$ we get, by monotone convergence,

$$\lim_{n\to\infty} \int_{K_n} |g(z)|^2 d\mu = \int_D |g(z)|^2 d\mu \leq c,$$

so $g \in L^2$.

Finally, by (3.2) $|f_n - g| \to 0$ uniformly on compacta of D and hence also pointwise. Moreover, $|f_n - g| \leq |f_n - f| + |f - g|$ pointwise on D. Now, $|f - g| \in L^2$ since both f and g are. Also, $|f_n - f|$ is certainly in L^2. Therefore so is the sum of these two functions. Since $|f_n - g|$ is dominated by an L^2 function, it follows that f_n converges to g in L^2. On the other hand, f_n also converges to f in L^2. Thus, $\| f - g \|_2 \leq \| f - f_n \|_2 + \| f_n - g \|_2$ each of which tends to zero and so $f = g$ a.e.

3.6 Spaces of holomorphic functions and Montel's theorem

Here we depart from our usual stance, i.e. assuming anything needed from real analysis, by redoing some very standard facts in order to bring the Ascoli-Arzela theorem into the picture.

Let Ω be a domain and $H(\Omega) \subseteq C(\Omega)$ be the subspace of holomorphic functions. We shall be interested in the topology of uniform convergence on compacta, $K \subseteq \Omega$, of functions both in $C(\Omega)$ and its subspaces. That is to say, convergence in $\|\|_K$, for each fixed K. Let n

be an integer and d be the usual distance in $\mathbb{C}$. We define

$$K_n = \left\{ z \in \mathbb{C} : |z| \leq n, d(z, \mathbb{C} - \Omega) \geq \frac{1}{n} \right\}.$$

Definition 3.6.1 We denote by $d(z, A)$, the distance from a point $z \in \mathbb{C}$ to a closed set $A \subseteq \mathbb{C}$ to be the $\inf\{d(z, a) : a \in A\}$. Moreover when we have a compact set $K \subseteq \mathbb{C}$, then $d(K, A) = \inf\{d(z, A) : z \in K\}$.

Lemma 3.6.2 *K_n is an increasing family of compact subsets of Ω whose union is all of Ω. Moreover, if K is any compact set in Ω, then $K \subseteq K_n$ for some n.*

Proof. If F is a closed subset of $\mathbb{C}$, then $z \mapsto d(z, F)$ is a continuous function. It follows that each K_n is a closed subset of $D(0, n)$ and hence is compact. Clearly, $K_n \subseteq K_{n+1}$. If $z \in K_n \cap \mathbb{C} - \Omega$, then $d(z, w) \geq \frac{1}{n}$ for all $w \in \mathbb{C} - \Omega$. But since $d(z, z) = 0$ this is a contradiction. Hence $K_n \subseteq \Omega$. Finally, if $z \in \Omega$, but not in any K_n, then there is a sequence $w_n \to z$, where $w_n \in \mathbb{C} - \Omega$. But then, $z \in \mathbb{C} - \Omega$ since this set is closed. This contradiction shows $\Omega = \cup_{n=1}^{\infty} K_n$.

If K is a compact subset of Ω then $d(K, \mathbb{C} - \Omega) > 0$. Choose n_1 large enough so that $d(K, \mathbb{C} - \Omega) \geq \frac{1}{n_1}$. Then $d(k, \mathbb{C} - \Omega) \geq \frac{1}{n_1}$ for all $k \in K$. Choose n_2 large enough so that $K \subseteq D(0, n_2)$. Then, taking n to be the maximum of these integers, we get $K \subseteq K_n$. $\square$

Definition 3.6.3 For f and $g \in C(\Omega)$ we define d^* by

$$d^*(f, g) = \sum_{n=1}^{\infty} \frac{1}{2^n} \cdot \frac{\| f - g \|_{K_n}}{(1 + \| f - g \|_{K_n})}.$$

Since $\frac{\|f-g\|_{K_n}}{1+\|f-g\|_{K_n}} \leq 1$ we see by the formula for the sum of a geometric series that $d^*(f, g) \leq \sum_{n=1}^{\infty} \frac{1}{2^n} < \infty$, so d^* is well defined.

Proposition 3.6.4 *$(C(\Omega), d^*)$ is a complete metric space. If f_n and f are in $C(\Omega)$, then f_n converges to f uniformly on compacta if and only if f_n converges to f in d^*.*

Proof. Clearly, $d^*(f,g) \geq 0$ and $d^*(f,g) = d^*(g,f)$. If $d^*(f,g) = 0$, then $\| f - g \|_{K_n} = 0$ for every n, that is, $f = g$ on K_n. Since the K_n fill out Ω, $f = g$.

To establish the triangle inequality it is clearly sufficient to see that

$$\frac{\| f - h \|_{K_n}}{1 + \| f - h \|_{K_n}} \leq \frac{\| f - g \|_{K_n}}{1 + \| f - g \|_{K_n}} + \frac{\| g - h \|_{K_n}}{1 + \| g - h \|_{K_n}}.$$

Let x and y be positive numbers. Then since $\frac{x}{1+x+y} \leq \frac{x}{1+x}$ and $\frac{y}{1+x+y} \leq \frac{y}{1+y}$ it follows that $\frac{x+y}{1+x+y} \leq \frac{x}{1+x} + \frac{y}{1+y}$ Therefore we need to know

$$\frac{\| f - h \|_{K_n}}{1 + \| f - h \|_{K_n}} \leq \frac{\| f - g \|_{K_n} + \| g - h \|_{K_n}}{1 + \| f - g \|_{K_n} + \| g - h \|_{K_n}}.$$

So it remains to prove that if a, b and c are positive numbers with $a \leq b + c$, then $\frac{a}{1+a} \leq \frac{b+c}{1+b+c}$. The easy verification of this inequality is left to the reader. Thus we have a metric space.

It follows from the nature of the metric d^* that if f_j converges to f in d^*, then f_j converges to f uniformly on each K_n. Since, if K is a compactum in Ω, then K is contained in K_n for some n. Hence we see that f_j converges to f uniformly on compacta. Conversely, suppose this is true for all compacta K. Then it is certainly true for all K_n, hence for all n, $\frac{\|f_j - f\|_{K_n}}{1+\|f_j - f\|_{K_n}}$ tends to zero. Let $\epsilon > 0$ and choose N so that $\sum_{n=N+1}^{\infty} \frac{1}{2^n} < \epsilon$. Choose j_0 so that $\frac{\|f_j - f\|_{K_n}}{1+\|f_j - f\|_{K_n}} < \epsilon$, for all $j \geq j_0$. If $n \leq N$, we know $K_n \subseteq K_N$ and hence $\| f_j - f \|_{K_n} \leq \| f_j - f \|_{K_N}$. Since as is easily verified for positive numbers, if $a \leq b$, then $\frac{a}{1+a} \leq \frac{b}{1+b}$, this gives

$$\sum_{n=1}^{N} \frac{1}{2^n} \frac{\| f_j - f \|_{K_n}}{1 + \| f_j - f \|_{K_n}} \leq \sum_{n=1}^{N} \frac{\epsilon}{2^n} \leq \sum_{n=1}^{\infty} \frac{\epsilon}{2^n} = \epsilon,$$

and

$$\sum_{N+1}^{\infty} \frac{1}{2^n} \frac{\| f_j - f \|_{K_n}}{1 + \| f_j - f \|_{K_n}} \leq \sum_{N+1}^{\infty} \frac{1}{2^n} < \epsilon.$$

Hence, $d^*(f_j, f) < 2\epsilon$ if $j \geq j_0$.

Finally, we prove the completeness of $(C(\Omega), d^*)$. Suppose there is some sequence f_j for which $d^*(f_j, f_k)$ tends to zero as j and k tend to ∞. Then, as we just saw, for each compact set K in Ω, $\| f_j - f_k \|_K$ tends to zero. In particular, this occurs at points. Hence, for each $z \in \Omega$, $|f_j(z) - f_k(z)|$ tends to zero as j and k go to ∞. By completeness of $\mathbb{C}$ there is a function f such that f_j converges to f pointwise. We show it converges uniformly on compacta. If we succeed, then by taking closed disks and using the fact that continuity is a local property, we would find that f continuous. Let K be a compact subset of Ω. Then for $z \in K$,

$$\begin{aligned} |f_j(z) - f(z)| &\leq |f_j(z) - f_k(z)| + |f_k(z) - f(z)| \\ &\leq \| f_j - f_k \|_K + |f_k(z) - f(z)| . \end{aligned}$$

For z fixed, choose k large enough so that $|f_k(z) - f(z)| < \epsilon$. Then choose j and k large enough so that $\| f_j - f_k \|_K < \epsilon$. Hence, for j large and $k \in K$, $|f_j(z) - f(z)| < 2\epsilon$. Therefore, $\| f_j - f \|_K \leq 2\epsilon$. Since this is true for any K by what we did just above, this means f_j converges to f in d^*. □

From Corollary 3.3.7 we arrive at:

Corollary 3.6.5 *$H(\Omega)$ as a closed subspace of $(C(\Omega), d^*)$ is also a complete metric space.*

Definition 3.6.6 A subset $\mathcal{F}$ of $H(\Omega)$ is called a *normal* family if its closure $\bar{\mathcal{F}}$ is compact.

Theorem 3.6.7 *A subset $\mathcal{F}$ of $H(\Omega)$ is a normal family if and only if it is locally bounded.*

Proof. The local boundedness condition is clearly equivalent to uniform boundedness on each closed disk in Ω and implies that for K, a point, $\mathcal{F}$ is pointwise bounded at each point of Ω. Moreover, by Corollary 3.3.12 we know $\mathcal{F}$ is also equicontinuous at each point. Therefore, by the Ascoli-Arzela theorem $\bar{\mathcal{F}}$ is compact [9, p. 155].

Conversely, suppose $\bar{\mathcal{F}}$ is compact. Consider $f \mapsto f|K$. This map is continuous since convergence in d^* implies uniform convergence on compacta. Therefore $\{f|K : f \in \bar{\mathcal{F}}\}$ is compact. Hence, $\bar{\mathcal{F}}$ is uniformly bounded on K and, therefore, so is $\mathcal{F}$. $\square$

3.7 The maximum modulus theorem and Schwarz' lemma

We first consider the maximum modulus theorem in its most basic form.

Theorem 3.7.1 *Let f be a holomorphic function on a domain Ω and suppose for some $a \in \Omega$ that $|f(a)| \geq |f(z)|$ for all $z \in \Omega$. Then f must be constant.*

Proof. Let $r > 0$ and γ_r be a small positively oriented circle centered at a and contained in the domain. By Corollary 3.2.2, $f(a) = \frac{1}{2\pi}\int_0^{2\pi} f(a+re^{it})dt$. Hence

$$|f(a)| \leq \frac{1}{2\pi}\left|\int_0^{2\pi} f(a+re^{it})dt\right| \leq \frac{1}{2\pi}\int_0^{2\pi} |f(a)|dt = |f(a)|\,,$$

since $|f(a+re^{it})| \leq |f(a)|$ for all t. Therefore, $\frac{1}{2\pi}\int_0^{2\pi} |f(a+re^{it})|dt \leq \frac{1}{2\pi}\int_0^{2\pi} |f(a)|dt$. Or alternatively,

$$\frac{1}{2\pi}\int_0^{2\pi} (|f(a)| - |f(a+re^{it})|)dt = 0\,.$$

But since $|f(a)| \geq |f(a+re^{it})|$ for all t and we are dealing with Riemann integration of continuous functions, this means $|f(a)| = |f(a+re^{it})|$ for all t and also for all small $r > 0$. Thus $|f|$ is constant on $D(a,r)$. But then, by Corollary 1.5.6, f itself must be constant on $D(a,r)$. By the identity theorem f is constant on Ω. $\square$

This immediate leads to the following:

Corollary 3.7.2 *For a non-constant holomorphic function f on domain Ω, $|f|$ can have no local maximum.*

Proof. Suppose $a \in \Omega$ were a local maximum point for $|f|$. Then there is an open disk in Ω centered at a where $|f(a)| \geq |f(z)|$. By Theorem 3.7.1 f is constant on this disk and therefore by the identity theorem f is constant on Ω. □

As mentioned above, in the case of a bounded domain Ω we know $\partial\Omega$ is non-trivial and $\bar{\Omega} = \Omega \cup \partial\Omega$.

Corollary 3.7.3 *Let Ω be a bounded domain and f a non-constant holomorphic function on Ω which is continuous on $\partial(\Omega)$. Then*

$$\max\{|f(z)| : z \in \bar{\Omega}\} = \max\{|f(z)| : z \in \partial(\Omega)\}.$$

Proof. The above statement means that $|f|$, which must have a maximum on the compact set $\bar{\Omega}$, actually assumes this maximum on the boundary. Since f is non-constant, this must be so for otherwise $|f|$ would assume its maximum on the interior, Ω contradicting Theorem 3.7.1. □

We can now apply this to study the behavior of entire functions at infinity. For example, $|e^z| = e^{\Re z} \to \infty$, as $\Re z \to \infty$. Also,

$$\sin z = \sin(x+iy) = \sin x \cos iy + \cos x \sin iy = \sin x \cosh y + i \cos x \sinh y.$$

Therefore, $|\sin z|^2 = \sin^2 x \cosh^2 y + \cos^2 x \sinh^2 y$. As $\Im z \to \infty$, so does $|\sin z|$.

This suggests that there is always some sequence tending to infinity along which f also tends to infinity.

Corollary 3.7.4 *Let f be a non-constant entire function. Then there exists a sequence $z_n \in \mathbb{C}$ tending to infinity such that $|f(z_n)| \to \infty$. In particular, the real or imaginary parts of such a function tend to $\pm\infty$.*

Proof. For $r > 0$, let $M(r) = \max_{|z| \leq r} |f(z)|$. (Since closed disks are compact the max is assumed.) By Corollary 3.7.3, $M(r) = \max_{|z|=r} |f(z)|$. $M(r)$ is non-negative and increasing with r so either $M(r)$ is bounded, or it tends to ∞. In the former case it means

f is a bounded function so, by Liouville's theorem, f would be constant. Hence the only possibility is $M(r)$ tends to ∞ as r does. This means there exists a sequence $z_n \in \mathbb{C}$ tending to infinity such that $|f(z_n)| \to \infty$. $\square$

Corollary 3.7.5 *Let u be a harmonic function defined on all of $\mathbb{R}^2$. If on $\mathbb{R}^2$, $u \leq c$, or, if $u \geq c$ where c is a constant, then u is itself constant.*

Proof. First suppose $u \leq c$. Since $\mathbb{C}$ is simply connected, by Corollary 2.5.4, $u = \Re(f)$, where f is entire. Because $f = u + iv$ we get $|e^{f(z)}| = e^{u(z)} \leq e^c$, another constant. Since $e^{f(z)}$ is entire and bounded it must be constant by Liouville's theorem. Therefore the derivative $\frac{d}{dz}e^{f(z)} = e^{f(z)}f'(z) = 0$ everywhere and since $e^{f(z)}$ is never zero we get $f'(z) = 0$ everywhere. Thus f and hence also u are constant. If $u \geq c$, then we still get $u = \Re(f)$, where f is entire. But then $-f(z) = -u(z) + iv(z)$ for some v. Since $u \geq c$ we have $-u \leq -c$. Applying the same reasoning, we find $-u$ is constant. Hence so is u. $\square$

We conclude this section with the Schwarz lemma. Here D denotes the unit disk centered at zero.

Proposition 3.7.6 *Let f be a holomorphic function on D satisfying $f(0) = 0$ and $|f(z)| \leq 1$ on D. Then*

(i) *$|f(z)| \leq |z|$ on D.*
(ii) *If $|f(a)| = |a|$ for some $a \neq 0 \in D$, then f is a rotation. (Thus, either $|f(z)| < |z|$ on D or $|f(z)| = |z|$ on D.)*
(iii) *$|f'(0)| \leq 1$.*
(iv) *If $|f'(0)| = 1$, then again f is a rotation.*

Proof. Since $f(0) = 0$, expanding f in a power series about 0 we get

$$f(z) = a_1 z + a_2 z^2 + \ldots, z \in D.$$

For $z \in D - (0)$, let $g(z) = \frac{f(z)}{z} = a_1 + a_2 z + \ldots$. We define $g(0) = a_1$. Now g is holomorphic on D. (We say g has a removable singularity at

$z = 0$.) Consider

$$\frac{g(z) - g(0)}{z} = a_2 + a_3 z + \cdots .$$

Taking the limit as $z \to 0$ gives a_2. Thus g is indeed holomorphic on D and $g'(0) = a_2$.

For $1 > r > 0$ let γ_r be a circle of radius r, centered at 0. If z lies on γ_r, then $|g(z)| = |\frac{f(z)}{z}| \leq \frac{1}{r}$. By the Maximum Modulus Theorem,

$$\max_{|z| \leq r} |g(z)| = \max_{|z|=r} |g(z)| .$$

Therefore, for all such r,

$$\max_{|z| \leq r} |g(z)| \leq \frac{1}{r} .$$

Letting $r \to 1$, we get $\max_{|z| \leq r} |g(z)| \leq 1$ on all of D. That is, $|f(z)| \leq |z|$ on D, proving (i).

Now suppose $|f(a)| = |a|$ for some $a \neq 0 \in D$. Then $|g(a)| = 1$. This means that $|g|$ achieves its maximum at a. Since a lies in the interior of our domain this is impossible unless g is constant on D. Therefore $g(z)$ must be constant and so from its power series we see that $a_j = 0$ for $j \geq 2$. Hence $f(z) = a_1 z$ on D. Also, since g is constant, $g(z) = a_1$ for all $z \in D$. But g has a maximum modulus value of 1 at a. Hence $|a_1| = 1$ and f is a rotation.

From the power series for f we see $f'(0) = a_1 = g(0)$. Since $\max_{|z| \leq r} |g(z)| \leq 1$ on all of D we see $|f'(0)| \leq 1$. Finally, suppose $|f'(0)| = 1$. Then $|g(0)| = 1$. This again contradicts the maximum modulus principle and so g would have to be constant. Reasoning as in the paragraph above, we again conclude f is a rotation. $\square$

Exercise 3.5 1. *Let $\{a_1, \ldots a_n\}$ be points in an open disk D and for $z \in D$ let*

$$d(z) = |z - a_1| \ldots |z - a_n|.$$

Does $d(z)$ have a maximum value on D?

2. *Let f be a non-constant holomorphic function on a bounded domain Ω and suppose f* is never zero on Ω. *Show $|f|$ cannot attain a minimum on Ω. (This is called the minimum modulus theorem.)*

3. *Explain what could happen if f is actually zero somewhere on Ω.*

4. *Suppose f is a non-constant holomorphic function on Ω and continuous on the boundary, $\partial(\Omega)$. Show if $|f|$ is constant on $\partial(\Omega)$, then f must have a zero in Ω.*

Chapter 4

Singularities

4.1 Classification of isolated singularities, the theorems of Riemann and Casorati-Weierstrass

Let f be a holomorphic function on a domain Ω and a be a point in Ω, where $f(a) = 0$. Then, expanding in a power series about a, we get $f(z) = \sum_{n=1}^{\infty} a_n(z-a)^n$. Then, of course, $a_0 = 0$. However, it may be that several more of the a_n beyond a_0 are equal to zero. But they can not all be zero, if $f \neq 0$.

Definition 4.1.1 The order of the zero, a, is the smallest positive integer, m, for which $a_m \neq 0$. Alternatively, the order of a is m means that $f(a) = \cdots = f^{m-1}(a) = 0$, but $f^m(a) \neq 0$. If the order of a zero is 1, then we say it is a *simple* zero.

This leads to the following:

Lemma 4.1.2 *If f has a zero of order n, we can write $f(z) = (z - a)^n g(z)$, where g is a holomorphic function with $g(a) \neq 0$.*

Definition 4.1.3 Here, instead of considering holomorphic functions, we will consider functions which are holomorphic on a domain Ω except for a finite (or discrete) set of points and at these points we have poles (see the definition below). Of course, if Ω is a bounded domain, such as the interior of a closed curve, then we are really talking about a finite set of points. Such functions are called *meromorphic* functions and the points are called the isolated singularities of f on Ω. By a discrete set of points, here we mean a subset of $\mathbb{C}$ having no limit points.

We will now give some examples of isolated singularities.

1. Let $\Omega = \mathbb{C}$ and $f(z) = \frac{1}{(z-a)^n}$, where $n \geq 1$. Then f has an isolated singularity at $z = a$.

2. More generally, let $f(z) = \frac{p(z)}{q(z)}$, where p and q are polynomials with degree $q \geq 1$ and the greatest common divisor of p and q is 1. That is to say, assume we have cancelled out any common factors of p and q. Then f has isolated singularities at the distinct roots of q (which exist by the Fundamental Theorem of Algebra).

3. A similar example is provided by $\tan z = \frac{\sin z}{\cos z}$. Here we have infinitely many, but discretely deployed singularities. These are the roots of cos. As an exercise the student should show these roots are all real and are therefore the usual ones for cos. We remark it is a theorem of Mittag-Leffler that, in any domain, each meromorphic function is a quotient $\frac{f(z)}{g(z)}$ of holomorphic functions exactly in this way.

4. Let $f(z) = \frac{\sin z}{z}$, if $z \neq 0$ and $f(0) = \alpha$, to be chosen later. Then 0 is certainly an isolated singularity of f.

Definition 4.1.4 An isolated singularity of f at $z = a$ is called *removable* if we can redefine f at a creating a new function g on the disk, $D(a, r)$ centered at a of radius $r > 0$, which is holomorphic at a and agrees with f on $D - \{a\}$.

The isolated singularity at 0 of $f(z) = \frac{\sin z}{z}$ is removable. Just take $g(0) = 1$. This choice is forced on us by the fact that $\lim_{z \to 0} \frac{\sin z}{z} = 1$. Therefore, if g is to be holomorphic at 0, it must be continuous there and so $g(0)$ must be 1. But then, $g(z) = 1 - \frac{z^2}{3!} + \frac{z^4}{5!} \dots$, for all $z \in \mathbb{C}$.

Since g is given by a convergent power series it is entire and in particular is holomorphic at 0. Clearly f and g coincide off 0.

Similarly, $\frac{e^z-1}{z}$ has a removable singularity at 0, by taking $g(0) = 1$.

On the other hand, if $n \geq 1$ then the isolated singularity of $f(z) = \frac{1}{(z-a)^n}$ at $z = a$ is not removable. This is because $|\frac{1}{(z-a)^n}|$ tends to ∞, as $z \to a$. Thus f is not bounded in any neighborhood of a and so can not even be extended to a continuous function at a.

This brings us to the Riemann removable singularities theorem.

Theorem 4.1.5 *Let f be a meromorphic function on Ω with a singularity at a. Then a is removable if and only if for some closed disk $\bar{D}$ centered at a, $\bar{D} \subseteq \Omega$, we have $\| f \|_{\bar{D}} < \infty$.*

Proof. If a is a removable singularity, then after redefining f at a we get a holomorphic and therefore continuous function defined in a neighborhood of a in Ω. Taking a suitable closed disk in $\bar{D}$ centered at a in Ω, we see f is bounded on $\bar{D}$ by compactness.

Conversely, if this is so for $z \in \Omega$, let $g(z) = (z-a)^2 f(z)$. Then g is holomorphic at a with $g'(a) = 0$. For

$$\frac{g(z) - g(a)}{z - a} = \frac{(z-a)^2 f(z)}{z - a} = (z-a)f(z) .$$

Since f is bounded in a neighborhood of a, $\lim_{z \to a}(z-a)f(z) = 0$. Thus $g'(a) = 0$. Because g is holomorphic at a it is continuous there. Hence, by the same reasoning,

$$g(a) = \lim_{z \to a}(z-a)^2 f(z) = 0 .$$

Therefore, in some neighborhood of a we know

$$g(z) = a_2(z-a)^2 + a_3(z-a)^3 + \cdots .$$

This means that near a we get

$$f(z) = \frac{g(z)}{(z-a)^2} = a_2 + a_3(z-a)^1 + \cdots .$$

It follows that a is a removable singularity of f and we take $f(a) = a_2 = \frac{g^{(2)}(a)}{2!}$. □

Now let us consider the only remaining possibility for an isolated singularity at a, namely that f is not bounded in any neighborhood of a. Nevertheless, it is possible that for some positive integer n, $(z-a)^n f(z) = g(z)$ is bounded in some neighborhood of a. Let n be the largest such positive integer (if there is a largest). Then, by the Riemann removable singularities theorem, g has a removable singularity at a. Therefore, $g(z) = \sum_{j=0}^{\infty} a_j (z-a)^j$. Hence for $z \in \Omega$,

$$f(z) = \frac{g(z)}{(z-a)^n} = \frac{a_0}{(z-a)^n} + \frac{a_1}{(z-a)^{n-1}} + \cdots + a_n + a_{n+1}(z-a) + \cdots .$$

When this happens, we say f has a *pole* at a of order n. For example, $f(z) = \frac{1}{(z-a)^n}$ has a pole at a of order n, since $(z-a)^n \frac{1}{(z-a)^n}$ is bounded in a neighborhood of a, but multiplying by $z-a$ to any smaller exponent gives something unbounded as $z \to a$.

Evidently, the only remaining possibility for an isolated singularity at a, is that for all non-negative integers n, $(z-a)^n f(z)$ is unbounded in any neighborhood of a. When this occurs, we say f has an essential isolated singularity at a.

For example, $\exp(\frac{1}{z})$ has an essential isolated singularity at 0. Clearly 0 is isolated. To see that it is essential we observe, for $z \neq 0$ that $\exp(\frac{1}{z}) = \sum_{n=0}^{\infty} \frac{1}{n! z^n}$. Clearly, no multiple by z^k, including $k = 0$, will make this bounded in a neighborhood of 0. Therefore, 0 is neither a pole nor a removable singularity. Obviously, similar remarks apply to $f(\frac{1}{z})$, where f is any holomorphic function in a neighborhood of 0, which is not a polynomial.

Exercise 4.1 *Notice, also, that the singularity at* 0 *of* $\exp(\frac{1}{z})$ *has some unusual features and we ask the student to verify these. As* $z \to 0$ *along the positive real axis,* $f(z) \to \infty$*, whereas, if* $z \to 0$ *along the negative real axis,* $f(z) \to 0$*. If* $z \to 0$ *along the imaginary axis, we can get as a limit any point on the unit circle. Finally, if* $z \to 0$ *along lines through the origin which are neither axis, we can get as a limit any point in* $\mathbb{C}$ *at all.*

These observations lead us quite directly to the theorem of Casorati-Weierstrass.

Theorem 4.1.6 *Let a be an isolated essential singularity of a meromorphic function f and $c \in \mathbb{C}$. Then there exists a sequence of distinct points $z_n \to a$ such that $f(z_n) \to c$.*

Proof. Suppose not. Then there exists an $\epsilon > 0$ so that $|f(z)-c| \geq \epsilon$, no matter how close z gets to a. In particular, $f(z)-c \neq 0$ for all z in some neighborhood of a. Let $g(z) = \frac{1}{f(z)-c}$, for z in this neighborhood. Then $|g(z)| = \frac{1}{|f(z)-c|} \leq \frac{1}{\epsilon}$. Hence $g(z)$ is bounded on this neighborhood. By Riemann's theorem g has a removable singularity at a. So

$$g(z) = a_0 + a_1(z-a) + a_2(z-a)^2 + \cdots.$$

Since $g(a)$ may be zero, we get

$$g(z) = a_n(z-a)^n + \cdots +,$$

where n is the order of the zero, i.e. $a_n \neq 0$. This means that for all such z we have

$$\frac{1}{g(z)} = \frac{1}{(z-a)^n} \frac{1}{\psi(z)},$$

where $\psi(a) \neq 0$. Hence, $\frac{1}{g(z)}$ is either holomorphic in a neighborhood of a, or has a pole. But $f(z) - c = \frac{1}{g(z)}$, for all such z. Hence $f(z) - c$ has a removable singularity or a pole at a. Of course, the same is then true of f itself. This is a contradiction. □

Corollary 4.1.7 *A non-constant entire function always has dense range.*

Proof. There are two possibilities: If f is a polynomial, let $w \in \mathbb{C}$. The equation $f(z) - w = 0$ being a polynomial equation, for a non-constant polynomial, must have a zero. Therefore, $f(z) = w$ for some z. Thus in this case, f maps onto $\mathbb{C}$. Now suppose f is not a polynomial. Let $g(z) = f(\frac{1}{z})$. Then g has an isolated singularity at $z = 0$ which is essential. By the Casorati-Weierstrass Theorem 4.1.6, if w is any point in $\mathbb{C}$, there is a sequence $z_n \to 0$ such that $f(\frac{1}{z_n}) \to w$. □

Exercise 4.2 *Prove this form of L'Hôpital's rule in the complex domain. Let f and g be non-constant holomorphic functions on a domain Ω and suppose for some point $a \in \Omega$, $f(a) = 0 = g(a)$. Prove that*

$$\lim_{z\to a}\frac{f(z)}{g(z)} = \lim_{z\to a}\frac{f'(z)}{g'(z)}.$$

4.2 The principle of the argument

In this section domains need not be simply connected.

Lemma 4.2.1 *Let f be a meromorphic (respectively holomorphic) function on a domain Ω and a be a point in Ω, where f has a zero of order n. Then there exists a meromorphic function (respectively holomorphic) g on Ω, with $g(a) \neq 0$ and*

$$\frac{f'(z)}{f(z)} = \frac{n}{z-a} + \frac{g'(z)}{g(z)},$$

for $z \in \Omega - \{a\}$.

Proof. Since f has a zero of order n we write $f(z) = (z-a)^n g(z)$, where $g(a) \neq 0$. Hence $f'(z) = n(z-a)^{n-1}g(z) + (z-a)^n g'(z)$. Therefore for $z \neq a$.

$$\frac{f'(z)}{f(z)} = \frac{n(z-a)^{n-1}g(z)}{(z-a)^n g(z)} + \frac{(z-a)^n g'(z)}{(z-a)^n g(z)},$$

and so

$$\frac{f'(z)}{f(z)} = \frac{n}{z-a} + \frac{g'(z)}{g(z)}.$$ □

The quantity $\frac{f'(z)}{f(z)}$, which we have also encountered earlier, is called the *logarithmic derivative* of f.

Applying the reasoning of Lemma 4.2.1 to the case of a pole of order m at a point $p \in \Omega$, we have a similar situation, namely, $f(z) = (z-p)^{-m}g(z)$, except the exponent is now negative. This leads to the following:

Lemma 4.2.2 *If f is a meromorphic function on Ω and p is a pole of order m, then there exists a meromorphic function g on Ω, with $g(p) \neq 0$ and*

$$\frac{f'(z)}{f(z)} = \frac{-m}{z-p} + \frac{g'(z)}{g(z)},$$

for $z \in \Omega - \{p\}$. Since $g(p) \neq 0$, its logarithmic derivative is holomorphic near p.

By use of Lemmas 4.2.1 and 4.2.2 several times we get the Principle of the argument, which provides an algebraic way of understanding $\frac{1}{2\pi i} \int_\gamma \frac{f'(z)}{f(z)} dz$. In particular, it tells us this quantity is always an integer.

Theorem 4.2.3 *Let f be a non-constant meromorphic function defined on a domain Ω, and γ be a positively oriented, piecewise smooth, simple closed curve in Ω such that f has no zeros or poles on the trajectory of γ. Then*

$$\frac{1}{2\pi i} \int_\gamma \frac{f'(z)}{f(z)} dz = Z_f - P_f,$$

where Z_f is the number of zeros of f within γ and P_f the number of poles of f within γ, each counted according to multiplicity.

Proof. Suppose inside γ, f has respectively k zeros $\{a_1, \dots, a_k\}$ of orders $\{n_1, \dots, n_k\}$ and l poles $\{p_1, \dots, p_l\}$ of orders $\{m_1, \dots, m_l\}$. Then

$$f(z) = \frac{(z-a_1)^{n_1} \dots (z-a_k)^{n_k}}{(z-p_1)^{m_1} \dots (z-p_l)^{m_l}} h(z),$$

where h is meromorphic and, just as f has no zeros or poles on the trajectory of γ, the same is true of h. Also, h has no zeros inside γ. Applying the two lemmas above several times we get

$$\frac{f'(z)}{f(z)} = \sum_{j=1}^{n} \frac{n_j}{z-a_j} + \sum_{l=1}^{m} \frac{-m_l}{z-p_l} + \frac{h'(z)}{h(z)}.$$

Hence

$$\int_\gamma \frac{f'(z)}{f(z)}dz = \sum_{j=1}^{n} n_j \int_\gamma \frac{1}{z-a_j}dz + \sum_{l=1}^{m}(-m_l)\int_\gamma \frac{1}{z-p_l}dz + \int_\gamma \frac{h'(z)}{h(z)}dz\,.$$

Since the logarithmic derivative of h is holomorphic, the last integral is zero. By the residue theorem (which we will prove formally in the next section) applied to the domain, the interior of γ minus the union of non-overlaping small disks with bounding circles γ_s contained inside γ centered at the zeros and the poles of f, it follows that

$$\frac{1}{2\pi i}\int_\gamma \frac{f'(z)}{f(z)}dz = \frac{1}{2\pi i}\left[\sum_{j=1}^{n} n_j \int_{\gamma_j} \frac{1}{z-a_j}dz + \sum_{l=1}^{m}(-m_l)\int_{\gamma_l} \frac{1}{z-p_l}dz\right]$$

$$= \sum_{j=1}^{n} n_j - \sum_{l=1}^{m} m_l = Z_f - P_f\,. \qquad \square$$

In particular this leads us to the following:

Corollary 4.2.4 *Let f be a holomorphic function defined on a domain Ω and γ be a positively oriented, piecewise smooth, simple closed curve in Ω such that f has no zeros on the trajectory of γ. Then*

$$\frac{1}{2\pi i}\int_\gamma \frac{f'(z)}{f(z)}dz = Z_f,$$

where Z_f is the number of zeros of f within γ counted according to multiplicity.

A slight generalization of Theorem 4.2.3 is the following. Of course, if $g \equiv 1$, we get the theorem itself.

Corollary 4.2.5 *Let f be a non-constant meromorphic function defined on a domain Ω and γ be a positively oriented, piecewise smooth, simple closed curve in Ω such that f has no zeros or poles on the trajectory of γ. If g is a holomorphic function on Ω, then*

$$\frac{1}{2\pi i}\int_\gamma g(z)\frac{f'(z)}{f(z)}dz = \sum_{j=1}^{n} g(a_j)n_j - \sum_{l=1}^{m} g(p_l)m_l,$$

where $\{a_1, \dots, a_k\}$ *are the zeros of* f *of orders* $\{n_1, \dots, n_k\}$ *and* $\{p_1, \dots, p_l\}$ *the poles of orders* $\{m_1, \dots, m_l\}$*, all within* γ*.*

Proof. Reasoning as in Theorem 4.2.3, we have

$$g(z)\frac{f'(z)}{f(z)} = \sum_{j=1}^{n} \frac{n_j g(z)}{z - a_j} + \sum_{l=1}^{m} \frac{-m_l g(z)}{z - p_l} + g(z)\frac{h'(z)}{h(z)}.$$

Integrating and taking into account that $\int_\gamma g(z)\frac{h'(z)}{h(z)}dz = 0$ gives

$$\int_\gamma g(z)\frac{f'(z)}{f(z)}dz = \sum_{j=1}^{n} \int_{\gamma_j} \frac{n_j g(z)}{z - a_j}dz + \sum_{l=1}^{m} \int_{\gamma_l} \frac{-m_l g(z)}{z - p_l}dz.$$

Let z_s stand for a zero or a pole of f of order, say, ν_s, and γ_s be small circle centered at z_s. Since g can be expanded in a power series at $z = z_s$, we get

$$\int_{\gamma_s} \frac{\nu_s g(z)}{z - z_s}dz = \nu_s \int_{\gamma_s} \frac{\sum_{j=0}^{\infty} b_j (z - z_s)^j}{z - z_s}dz = \nu_s \int_{\gamma_s} \sum_{j=0}^{\infty} b_j (z - z_s)^{j-1} dz.$$

But, since $\int_{\gamma_s} \sum_{j=1}^{\infty} b_j (z - z_s)^{j-1} dz = 0$, and $b_0 = g(z_s)$, we get

$$\nu_s \int_{\gamma_s} b_0 \frac{1}{z - z_s}dz = 2\pi i g(z_s)\nu_s.$$

Hence, the integral calculation at the beginning gives

$$\frac{1}{2\pi i}\int_\gamma g(z)\frac{f'(z)}{f(z)}dz = \sum_{j=1}^{n} g(a_j)n_j - \sum_{l=1}^{m} g(p_l)m_l. \qquad \square$$

As we shall see later in the language of residues, what we have shown here is that

$$\operatorname{Res}\left(g\frac{f'}{f}, z_s\right) = g(z_s)\operatorname{Res}\left(\frac{f'}{f}, z_s\right).$$

We can use Corollary 4.2.5 to "explicitly" calculate the inverse of a 1:1 holomorphic function. Of course, by Corollary 1.4.7 we already know the inverse is holomorphic. Let f be a 1:1 holomorphic function defined on a domain containing a disk, $D(a, r_0)$, together with its boundary γ. Since f is 1:1, its derivative is $\neq 0$ everywhere on $D(a, r_0)$. By Corollary 1.4.8, $f(D(a, r_0)) = \Omega$ is open. It is the function $f : D(a, r_0) \to \Omega$ whose inverse we will calculate.

Corollary 4.2.6 *For* $w \in \Omega$, $f^{-1}(w) = \frac{1}{2\pi i} \int_\gamma \frac{zf'(z)}{f(z)-w} dz$.

Proof. Here $w \in \Omega$ is arbitrary, but fixed. Since $\frac{d}{dz}(f(z) - w) = f'(z)$ we see the integrand is really $\frac{z(f-w)'(z)}{f(z)-w}$. Taking the function $g(z) = z$ in the result above, we get

$$\frac{1}{2\pi i} \int_\gamma \frac{zf'(z)}{f(z) - w} dz = \sum_{j=1}^{n} a_j n_j \,,$$

where we have the zeros and their multiplicities of the function $f - w$ inside γ, because there are no poles. To know this, we must also check that the function $f - w$ has no zeros on γ. What are the zeros of this function both within and on γ? If $f(z) - w = 0$, then $f(z) = w$ and, since f is 1:1, $z = f^{-1}(w) \in D(a, r_0)$. Therefore, there are no zeros on γ and a unique zero inside γ. It has multiplicity 1 since the derivative is everywhere non-zero. Hence $\sum_{j=1}^{n} a_j n_j = f^{-1}(w)$. $\square$

Our final result in this section is a geometric way of understanding $\frac{1}{2\pi i} \int_\gamma \frac{f'(z)}{f(z)}$. This is also called the Principle of the Argument. It tells us that computing an integer one way is the same as computing it a different way.

Let f be a holomorphic function defined on a domain Ω and γ be a positively oriented, *smooth*, simple closed curve in Ω such that f has no zeros on the trajectory of γ. Now $f \circ \gamma$ is a smooth closed curve, but it may not be simple and it may not be positively oriented. (Of course, if f were 1:1, then it would be simple.) Notice that $f \circ \gamma$ never passes through zero since f has no zeros on the trajectory of γ. Choose a point w_0 on $f \circ \gamma$ and, since $w_0 \neq 0$, let $\theta_0 = \arg w_0$. Because $\arg w$

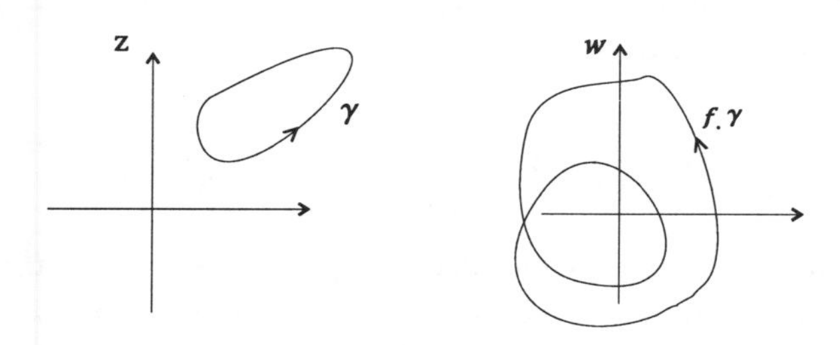

Figure 4.1

varies continuously as we go through our parameter domain, when we get back to w_0 we can ask how many times we have gone around the origin. We call this number $\frac{1}{2\pi}\Delta_\gamma f$. By definition it is an integer. It is clearly independent of the choice of w_0 (cf. Figure 4.1).

Theorem 4.2.7 *Let f be a holomorphic function defined on a domain Ω and γ be a positively oriented, smooth, simple closed curve in Ω such that f has no zeros on the trajectory of γ. Then*

$$\frac{1}{2\pi i}\int_\gamma \frac{f'(z)}{f(z)}dz = \frac{1}{2\pi}\Delta_\gamma f\,.$$

Proof. Let $w = f(z)$ and apply the change of variable formula for contour integrals (Proposition 2.2.5). Since $dw = f'(z)dz$, $\frac{dw}{w} = \frac{f'(z)}{f(z)}dz$, we get $\frac{1}{2\pi i}\int_\gamma \frac{f'(z)}{f(z)}dz = \frac{1}{2\pi i}\int_{f\circ\gamma}\frac{dw}{w}$. Now consider the parameterization of $\gamma : [a,b] \to \Omega$ and the transformed contour $w(t) = f(\gamma(t))$. Since f has no zeros on γ, writing $w(t) = \rho(t)e^{i\theta(t)}$ in polar coordinates and differentiating with respect to t gives

$$\frac{w'(t)}{w(t)} = \frac{\rho'(t)}{\rho(t)} + i\theta'(t)\,.$$

Hence,

$$\int_{f\circ\gamma}\frac{dw}{w} = \int_a^b \frac{w'(t)}{w(t)}dt = \int_a^b \frac{\rho'(t)}{\rho(t)}dt + i\int_a^b \theta'(t)dt\,.$$

But this is simply $\log \rho(t) + i\theta(t)$ evaluated between a and b. Since the curve $f \circ \gamma$ is closed, $\rho(a) = \rho(b)$. Also $\theta(b) - \theta(a) = \Delta_\gamma f$. Hence $\int_{f\circ\gamma} \frac{dw}{w} = i\Delta_\gamma f$, so that

$$\frac{1}{2\pi i}\int_\gamma \frac{f'(z)}{f(z)}dz = \frac{1}{2\pi i}\int_{f\circ\gamma} \frac{dw}{w} = \frac{1}{2\pi}\Delta_\gamma f. \qquad \square$$

Exercise 4.3 1. *For a positive integer N, let γ_N be the positively oriented square in the plane, centered at the origin with sides parallel to the coordinate axes of length $2N$. Calculate $\int_{\gamma_N} \tan z dz$.*

2. *Suppose $p(z) = a_m z^m + a_{m-1}z^{m-1} + \cdots + a_0$ and $q(z) = b_n z^n + b_{n-1}z^{n-1} + \cdots + b_0$ are polynomials with $\deg p = m$ and $\deg q = n$. Let γ_r be a positively oriented circle centered at zero of radius r. Show (a) if $n \geq m+2$, then $\lim_{r\to\infty} \int_{\gamma_r} \frac{p(z)}{q(z)}dz = 0$ and (b). if $n = m+1$, then $\lim_{r\to\infty} \int_{\gamma_r} \frac{p(z)}{q(z)}dz = 2\pi i \frac{a_m}{b_n}$.*

4.3 Rouché's theorem and its consequences

As above, for h a meromorphic function defined on a domain Ω and γ be a positively oriented, piecewise smooth, simple closed curve in Ω, we denote by Z_h the number of zeros of h within γ and P_h the number of poles of h within γ, each counted according to multiplicity.

Theorem 4.3.1 *Let Ω be a simply connected domain, γ be a positively oriented, piecewise smooth, simple closed curve in Ω, and f and g be meromorphic functions defined on Ω with the property that, on γ, $|f(z)| > |g(z)|$ and neither f nor g has zeros or poles on γ. Then $Z_{f+g} - P_{f+g} = Z_f - P_f$.*

Moral: The location and multiplicity of the zeros and of the poles inside γ may be changed by adding a "trivial" g, but $Z - P$ does not. For example, a double root (or pole) may be changed to two simple roots (or poles), or vice versa.

Our next corollary, which is also called Rouché's theorem, follows since in the case of holomorphic functions there are no poles.

Corollary 4.3.2 *Let Ω be a simply connected domain, γ be a positively oriented, piecewise smooth, simple closed curve in Ω and f and g be holomorphic functions defined on Ω such that on γ, $|f(z)| > |g(z)|$ and such that neither f nor g has zeros on γ. Then $Z_{f+g} = Z_f$.*

We remark that the condition $|f(z)| > |g(z)|$ cannot be weakened to $|f(z)| \geq |g(z)|$. For example, in the corollary above, if $g = -f$, then $|f(z)| = |g(z)|$ everywhere on Ω. But $f + g \equiv 0$ so the number of zeros of $f + g$ inside γ is infinite. However, if $f \neq 0$, the number of zeros of f inside γ is finite.

We now give two examples of specific polynomials showing how Rouché's theorem can be used to approximately locate the roots. Later we shall apply Rouché's theorem to general polynomials and other holomorphic functions. In the next section we shall apply it to transcendental meromorphic functions.

Example Suppose we wanted to know how many zeros $z^7 - 4z^3 + z - 1$ has inside the unit circle? Here $\Omega = \mathbb{C}$. Let $f(z) = -4z^3$ and $g(z) = z^7 + z - 1$. When $|z| = 1$ we see $|f(z)| = 4$, while $|g(z)| \leq 3$. So $|f(z)| > |g(z)|$ on the unit circle, γ. Hence the number of zeros of f inside γ, which is clearly 3, is the same as the number of zeros of $f + g = z^7 - 4z^3 + z - 1$.

As another example, we consider the polynomial $p(z) = z^3 + z + 1$. Here all roots of p have their modulus between $\frac{2}{3}$ and $\frac{4}{3}$. For let $f(z) = z^3$ and $g(z) = z+1$. On the circle, $|z| = \frac{4}{3}$ $|f(z)| = \frac{64}{27}$, while $|z+1| \leq |z|+1$. So $|g(z)| \leq \frac{63}{27}$. Therefore, on this circle $|f(z)| > |g(z)|$. Since f clearly has 3 roots inside this circle so does $f+g = p$. Therefore, all the roots of p are inside this circle. Now consider the circle $|z| = \frac{2}{3}$. Here $|f(z)| = \frac{8}{27}$, while $|1 + z| > \frac{1}{3} > \frac{8}{27}$. So here things are reversed and $|g(z)| > |f(z)|$. Therefore, within the smaller circle p has the same number of roots as g has. But the only root of g is $z = -1$ which lies outside this circle. Hence p has no roots inside $|z| = \frac{2}{3}$.

Before proving the theorem it will be convenient to formulate a lemma.

Lemma 4.3.3 *Let all quantities be as in the theorem and for* $0 \le t \le 1$ *we define*

$$\Phi_t(z) = f(z) + tg(z).$$

Then

(i) *Each* Φ_t *is meromorphic on* Ω.
(ii) $\Phi_0 = f$ *and* $\Phi_1 = f + g$.
(iii) *There is some positive number* m *such that* $|\Phi_t(z)| \ge m$ *everywhere on* γ.

Proof. The first two items are clear. To prove the third, observe that since for a and $b \in \mathbb{C}$, we know $|a - b| \ge ||a| - |b||$. It follows that

$$|f(z) - (-tg(z))| \ge |f(z)| - |tg(z)| = |f(z)| - t|g(z)|.$$

But since $0 \le t \le 1$ and $|g(z)| \ge 0$, $|f(z)| - t|g(z)| \ge |f(z)| - |g(z)|$. Therefore, $|\Phi_t(z)| = |f(z) + tg(z)| \ge |f(z)| - |g(z)| > 0$ on γ. Since $|f(z)| - |g(z)|$ is a continuous function on Ω, it attains its minimum value, $m > 0$, on the compact trajectory of γ. □

We now turn to the proof of Rouché's theorem. For $0 \le t \le 1$, let $h(t) = \frac{1}{2\pi i}\int_\gamma \frac{\Phi_t'(z)}{\Phi_t(z)}dz$. Now each Φ_t is meromorphic on a simply connected domain in which γ is a positively oriented, piecewise smooth, simple closed curve. By the principle of the argument (Theorem 4.2.3), h takes integer values. Namely, $h(t)$ is the number of zeros minus the number of poles of $\Phi_t(z)$ inside γ. We will show h is continuous. Therefore, since the unit interval is connected, h is constant. In particular, $Z(\Phi_0) - P(\Phi_0) = Z(\Phi_1) - P(\Phi_1)$. That is, $Z_f - P_f = Z_{f+g} - P_{f+g}$.

Now to see that h is continuous, let t_1 and $t_2 \in [0,1]$. Then

$$\begin{aligned} h(t_2) - h(t_1) &= \frac{1}{2\pi i}\int_\gamma \left(\frac{f' + t_2 g'}{f + t_2 g}\right) - \left(\frac{f' + t_1 g'}{f + t_1 g}\right) dz \\ &= \frac{t_2 - t_1}{2\pi i}\int_\gamma \frac{g'f - f'g}{(f + t_2 g)(f + t_1 g)} dz\,. \end{aligned}$$

Now let M be the maximum values of all the functions f, g, f' and g' on γ and m be the minimum value of $|f| - |g|$ on γ. As we saw, for all $t \in [0,1]$, and z on γ $|f + tg| \geq |f(z)| - |g(z)| \geq m$. It follows that $\frac{1}{|f+tg|} \leq \frac{1}{m}$, for all such t, z. Therefore,

$$|h(t_2) - h(t_1)| \leq \frac{|t_2 - t_1|}{2\pi} \frac{2M^2}{m^2} L(\gamma).$$

Since $\frac{M^2}{\pi m^2} L(\gamma)$ is a constant, h is continuous.

We conclude this section with various consequences of Rouché's theorem.

Our first is a new and sharper form of the Fundamental Theorem of Algebra. Because it is not a proof by contradiction, it gets all the roots not just some root, but more importantly it locates them in a disk.

Corollary 4.3.4 *If $p_n(z) = \sum_{j=0}^{n} a_j z^j$ is a polynomial of degree $n > 1$, then all its roots lie in the disk $D(0, r)$, where $r > \max\left\{\frac{\sum_{j=0}^{n-1} |a_j|}{|a_n|}, 1\right\}$.*

Proof. If $|z| = r > 1$ we have

$$\frac{|p_{n-1}(z)|}{|a_n z^n|} \leq \frac{\sum_{j=0}^{n-1} |a_j| \cdot r^{n-1}}{|a_n| r^n} = \frac{\sum_{j=0}^{n-1} |a_j|}{|a_n| r}.$$

Therefore, on this circle, $|p_{n-1}(z)| < |a_n z^n|$. By Rouché's theorem, p_n has the same number of zeros inside the circle as $a_n z^n$ does. This is clearly n. So all zeros of p_n lie in $D(0, r)$. □

Our next corollary is a fixed point theorem. Here we get a sharper result than one would get by using the Brouwer fixed point theorem since our fixed point lies in the interior, while the Brouwer theorem merely places it on the closed disk. Here the fixed point is also unique.

Corollary 4.3.5 *Let f be a holomorphic function on a domain Ω containing the open unit disk D, together with its boundary γ. If $|f(z)| < 1$ on γ, then f has a unique fixed point in D.*

Note that by the maximum modulus theorem

$$\max\{|f(z)| : |z| \leq 1\} = \max\{|f(z)| : |z| = 1\} < 1.$$

So that $f(\bar{D}) \subseteq D$. In particular, $f(D) \subseteq D$ and $f(\bar{D}) \subseteq \bar{D}$.

Proof. On γ we know $|f(z)| < 1 = |z|$. Therefore, by Rouché, $z - f(z)$ has the same number of zeros in D as z has, namely, one zero. Thus f has a unique fixed point on D. □

Our next result is known as Hürwitz' theorem.

Corollary 4.3.6 *Let $\{f_n\}$ be a sequence of holomorphic functions on a domain Ω converging to f uniformly on compacta and $\bar{D}$ be a closed disk in Ω with boundary γ. If f has no zeros on γ, then there is some index n_0 after which f_n and f have the same number of zeros in D.*

Before we proceed with the proof we remark that, if $f \not\equiv 0$, then given a point $a \in \Omega$ there is always such a disk centered at a. Otherwise, by taking radii tending to zero we would get a distinct sequence of points $z_n \to a$ with $f(z_n) = 0$ for all n. By the identity theorem, $f \equiv 0$.

Proof. Let $m > 0$ be the minimum value of $|f(z)|$ on γ. Since $\bar{D}$ is compact, $\{f_n\}$ converges to f uniformly on $\bar{D}$. Therefore there is some index n_0 after which $|f_n(z) - f(z)| < m$, for all $z \in \bar{D}$. But since $m \leq |f(z)|$, for z on γ, Rouché's theorem applies. For all $n \geq n_0$, $f_n = f_n - f + f$ and f have the same number of zeros in D. □

Finally, just for fun, we remark that Rouché's theorem also implies the maximum modulus theorem.

Corollary 4.3.7 *Let f be a holomorphic function on a domain Ω. If $|f(z)|$ has a maximum, it must take it on the boundary.*

Proof. Suppose not. Suppose $|f(z)|$ takes its maximum value at an interior point, a. Take a closed disk $\bar{D} \subseteq \Omega$ centered at a. Then $|f(a)| > |f(z)|$ for all $z \in \bar{D}$, $z \neq a$. In particular, this holds for all z on the boundary γ. Thus for the constant function, $-f(a)$, we see $|-f(a)| > |f(z)|$, for all z on γ. Therefore, the number of zeros of $-f(a)$ in D equals the number of zeros of $-f(a) + f(z)$ in D. But $-f(a) \neq 0$ since $|f(a)| > |f(z)| \geq 0$ for all z on γ. Hence $-f(a) + f(z)$ has no zeros in D. But $z = a$ is clearly such a zero. This is a contradiction. □

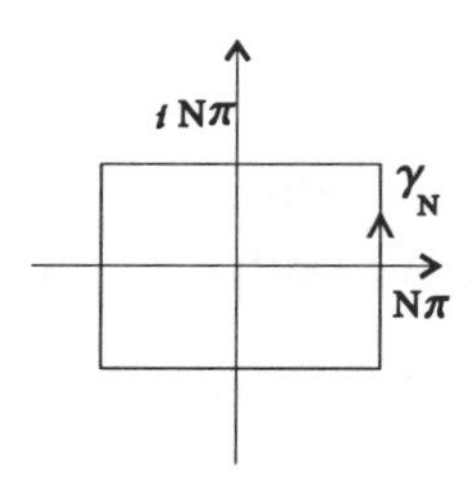

Figure 4.2

4.4 The study of a transcendental equation

Our final application of Rouché's theorem is the study of the transcendental equation,

$$\tan z = \lambda z\,,$$

where $z \in \mathbb{C}$ and $\lambda > 0$. This equation is of some importance in the study of heat flow.

Let γ_N be the positively oriented, piecewise smooth, simple closed curve whose trajectory is the sides of a square centered at the origin with sides parallel to the coordinate axis of distance $2\pi N$ from one another (cf. Figure 4.2).

Lemma 4.4.1 *Given $\epsilon > 0$ there is an N large enough so that $|\tan z| < 1 + \epsilon$ for all z on the trajectory of γ_N.*

Proof. For $z = x + iy$ we have by the identity theorem

$$\tan z = \tan(x + iy) = \frac{\tan x + \tan(iy)}{1 - \tan x \tan(iy)}\,.$$

But also

$$\tan(iy) = \frac{\sin(iy)}{\cos(iy)} = \frac{i\sinh(y)}{\cosh(y)} = i\tanh(y)\,.$$

Hence

$$\tan z = \frac{\tan x + i\tanh(y)}{1 - i\tan x\tanh(y)}\,.$$

Therefore

$$|\tan z|^2 = \frac{|\tan x + i\tanh y|^2}{|1 - i\tan x \tanh y|^2} = \frac{\tan^2 x + \tanh^2 y}{1 + \tan^2 x \tanh^2 y}\,.$$

First we consider the horizontal sides of γ_N; $y = \pm N\pi$, $|x| \le N\pi$. Since $|y| = N\pi$, as $N \to \infty$, $|y| \to \infty$. So that $\lim_{|y|\to\infty} |\tanh y| = 1$. Hence

$$\lim_{|y|\to\infty} |\tan z|^2 = \lim_{|y|\to\infty} \frac{\tan^2 x + \tanh^2 y}{1 + \tan^2 x \tanh^2 y} = 1\,.$$

Hence for N large enough $|\tan z|^2 < 1 + \epsilon$ on the horizontal sides.

On the vertical sides, $|x| = N\pi$ and $|y| \le N\pi$. If $x = 0$ and $|y| \le N\pi$ then $\tan z = \tan(iy) = i\tanh y$. So $|\tan(0 + iy)|^2 = \tanh^2 y \le 1$. But, by the identity theorem, tan is periodic of period π. Therefore, $\tan(\pm N\pi + iy) = \tan(iy)$ so $|\tan z|^2 \le 1$ on the vertical axis. Therefore, for N large enough $|\tan z| < 1 + \epsilon$ on the entire trajectory of γ_N. □

Corollary 4.4.2 *For N large enough $|\tan z| < |\lambda z|$ on the trajectory of γ_N.*

Proof. If we take $N \ge \frac{1+\epsilon}{\lambda\pi}$, then for z on the trajectory of γ_N we have

$$|\lambda z| = \lambda|z| \ge \lambda N\pi \ge 1 + \epsilon > |\tan z|\,.$$ □

Lemma 4.4.3 *The only zeros or poles of* $\tan z$ *lie on the* $\mathbb{R}$*-axis. In particular, they are isolated and therefore* $\tan z$ *is a meromorphic function.*

Proof. That is to say, the only zeros of $\sin z$ or $\cos z$ are real. It will be sufficient to prove this for $\cos z$ since by the identity theorem $\sin z = \cos(\frac{\pi}{2} - z)$. Now, also by the identity theorem, $\cos(x+iy) = \cos x \cosh y - i \sin x \sinh y$. If this is zero, then $\cos x = i \sin x \tanh y$. Since everything in sight is real, either $\sin x = 0$, or $\tanh y = 0$. But if $\sin x = 0$, then also $\cos x = 0$. This is impossible. Therefore, $\tanh y = 0$ and so $y = 0$ (cf. Figure 4.3). □

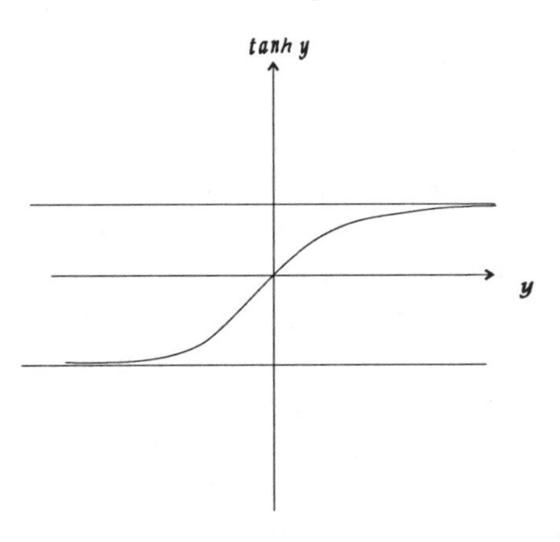

Figure 4.3

Corollary 4.4.4 *Consider the equation* $\tan z = \lambda z$, *where* $\lambda > 0$. *For* N *sufficiently large the number of roots (counting multiplicity) inside* γ_N *is* $2N + 1$.

Proof. By the corollary above, for N large enough $|\tan z| < |\lambda z|$ on the trajectory of γ_N. Since here we have a meromorphic function, by Rouché's theorem the number, Z, of zeros minus the number, P, of poles of $\tan z - \lambda z$ inside γ_N is equal to number of zeros minus the number of poles of λz inside γ_N, this latter number being $1 - 0$. Thus $Z(\tan z - \lambda z) - P(\tan z - \lambda z) = 1$. But $P(\tan z)$ inside $\gamma_N = Z(\cos z)$ inside γ_N. Here there are clearly $2N$ zeros all real. If $\cos x = 0$, then $\frac{d}{dx}(\cos x) = -\sin x \neq 0$. Thus all these roots are *simple*. It follows that inside γ_N counting multiplicity $Z(\tan z - \lambda z) = 2N + 1$. □

We remark that just because the poles of $\tan z - \lambda z$ are real and simple does not make the zeros either real, or simple. To analyze these zeros we consider three cases: $\lambda > 1$, $\lambda = 1$ and $\lambda < 1$. The first of these is the easiest. Inspection of the graph of $y = \tan x$ shows there are at least $3 + 2(N - 1) = 2N + 1$ real zeros of $\tan x - \lambda x$ in $[-N\pi, N\pi]$. Therefore, in this case, all the zeros inside γ_N are real and simple and taking $N \to \infty$ we see that when $\lambda > 1$ our equation has only real simple zeros. If $\lambda = 1$, then the equation is $\frac{\sin z}{\cos z} = z$. Expanding these functions in power series about $z = 0$ tells us $z^3(\frac{1}{2!} - \frac{1}{3!}) + z^5(\frac{1}{4!} - \frac{1}{5!}) + \cdots = 0$. Therefore $z = 0$ is a triple root. Arguing as in the previous case we get another $2(N-1)$ distinct roots giving a total of $3 + 2(N-1) = 2N+1$ real zeros of $\tan x - \lambda x$ in $[-N\pi, N\pi]$. Therefore, in this case, all the zeros inside γ_N are real and simple except 0 which has multiplicity 3. Taking $N \to \infty$ we see that when $\lambda = 1$ our equation has only real zeros all of which are simple except 0 which has multiplicity 3.

If $0 < \lambda < 1$, then $\tan z = \lambda z$ has 2 pure imaginary conjugate roots since $\tan(iy) = \lambda iy$ means $\tanh y = \lambda y$ and since $0 < \lambda < 1$ there are exactly 3 distinct y satisfying this equation, $y = 0$ and two others. Therefore, $z = 0$ is a simple root and $\pm y$ are two pure imaginary simple roots. Arguing as in the two previous cases, we get another $2(N-1)$ distinct roots giving a total of $3 + 2(N-1) = 2N+1$ zeros of $\tan z - \lambda z$ inside γ_N. Therefore, also in this case, taking $N \to \infty$ we see that our equation has only simple roots, two of which are pure imaginary and the rest all real (cf. Figures 4.4 and 4.5).

4.5 Laurent expansion

For $0 \le r_1 < r_2 \le \infty$, we consider the open annulus, $A(r_1, r_2)$ centered at a point $a \in \mathbb{C}$ (cf. Figure 4.6).

$$A(r_1, r_2) = \{z \in \mathbb{C} : r_1 < |z - a| < r_2\}.$$

The cases, $r_1 = 0$ and $r_2 = \infty$ are of interest here. They are the punctured disk and punctured plane.

Our task here is to categorize all holomorphic functions on $A(r_1, r_2)$. We first construct such functions. Let $\sum_{n=0}^{\infty} b_n \zeta^n$ be a convergent power series (holomorphic function) in a neighborhood of zero of radius $\frac{1}{r_1}$. Then $\sum_{n=0}^{\infty} b_n (z-a)^{-n}$ converges in the exterior of the circle centered at a of radius r_1. On the other hand, suppose $\sum_{n=0}^{\infty} c_n (z-a)^n$ converges for $|z - a| < r_2$. Then in $A(r_1, r_2)$ both series converge and, hence, so does their sum.

$$\sum_{n=0}^{\infty} b_n (z-a)^{-n} + \sum_{n=0}^{\infty} c_n (z-a)^n = \sum_{n=-\infty}^{\infty} a_n (z-a)^n .$$

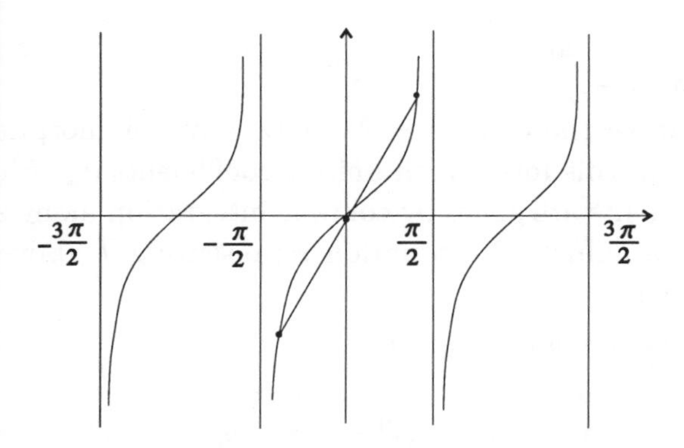

Figure 4.4

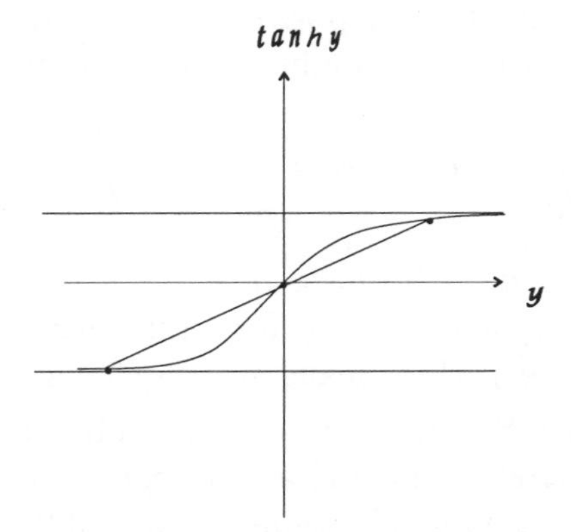

Figure 4.5

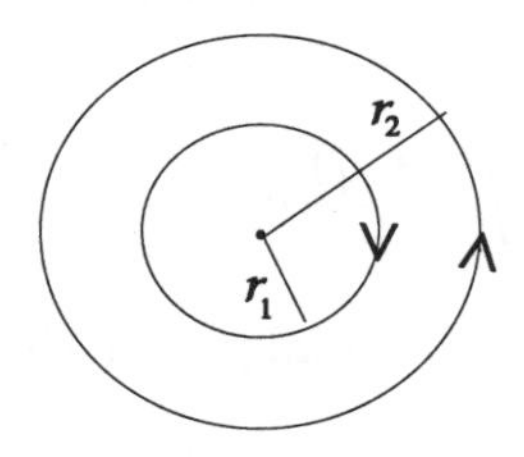

Figure 4.6

Here $b_0 + c_0 = a_0$ and $f(z) = \sum_{n=-\infty}^{\infty} a_n(z-a)^n$ is a holomorphic function on $A(r_1, r_2)$.

We now prove the converse. Namely, any holomorphic function f on $A(r_1, r_2)$ is of this form, with unique coefficients a_n. Moreover these Laurent coefficients are given by contour integration along simple closed curves in the annulus. The equation representing f above is called its Laurent expansion.

Proof of uniqueness. Suppose

$$\sum_{n=-\infty}^{\infty} a_n(z-a)^n = \sum_{n=-\infty}^{\infty} b_n(z-a)^n .$$

Multiply both sides by $(z-a)^j$, where $j \in \mathbb{Z}$ and then integrate both sides of this equation along a positively oriented, piecewise smooth, simple closed curve γ in the annulus which contains a in its interior.

$$\sum_{n=-\infty}^{\infty} a_n \int_\gamma (z-a)^{n+j} dz = \sum_{n=-\infty}^{\infty} b_n \int_\gamma (z-a)^{n+j} dz .$$

Since $\int_\gamma (z-a)^{n+j} dz = 0$ unless $n + j = -1$, in which case it gives $2\pi i$, we get $2\pi i a_{-(j+1)} = 2\pi i b_{-(j+1)}$ for all integers j. Since any integer $n = -(j+1)$ for some j, this means $a_n = b_n$ for all n.

Proof of existence. Suppose f is a holomorphic function on $A(r_1, r_2)$. Here we recall Cauchy's theorem and integral formula in the case of multiply connected domains (Sections 3.1 and 3.2). Let γ_1 and γ_2 be concentric circles, the first slightly larger and the second slightly smaller than the inner and outer boundary circles of $A(r_1, r_2)$. For any $z \in A(r_1, r_2)$ we can always choose such circles so that z lies between them. Then we get

$$f(z) = \frac{1}{2\pi i} \int_{\gamma_2} \frac{f(\zeta)}{\zeta - z} d\zeta - \frac{1}{2\pi i} \int_{\gamma_1} \frac{f(\zeta)}{\zeta - z} d\zeta .$$

We consider the integrals individually. For the first

$$\frac{1}{\zeta - z} = \left(\frac{1}{\zeta - a}\right)\left(\frac{1}{1 - \frac{z-a}{\zeta - a}}\right),$$

where ζ is on γ_2. Since $|\frac{z-a}{\zeta-a}| < 1$ we have a geometric series and get $\frac{1}{\zeta-z} = \frac{1}{\zeta-a}\sum_{n=0}^{\infty}(\frac{z-a}{\zeta-a})^n$. So

$$\frac{1}{\zeta - z} = \sum_{n=0}^{\infty} \frac{(z-a)^n}{(\zeta-a)^{n+1}}\,.$$

In the case of the second integral, where ζ is on γ_1, we get

$$\frac{1}{\zeta - z} = \left(\frac{-1}{z-a}\right)\left(\frac{1}{1-\frac{\zeta-a}{z-a}}\right).$$

Here $|\frac{\zeta-a}{z-a}| < 1$ and so we again get a geometric series, and

$$\frac{1}{\zeta - z} = -\sum_{n=0}^{\infty} \frac{(\zeta-a)^n}{(z-a)^{n+1}}\,.$$

Since in both cases the convergence is uniform on compacta, i.e. on the closed region between γ_1 and γ_2, just as in the Taylor series case we can integrate term by term. So

$$\begin{aligned} f(z) &= \frac{1}{2\pi i}\int_{\gamma_2} f(\zeta)\left(\sum_{n=0}^{\infty}\frac{(z-a)^n}{(\zeta-a)^{n+1}}\right)d\zeta \\ &\quad - \frac{1}{2\pi i}\int_{\gamma_1} f(\zeta)\left(-\sum_{n=0}^{\infty}\frac{(\zeta-a)^n}{(z-a)^{n+1}}\right)d\zeta \\ &= \sum_{n=0}^{\infty}(z-a)^n \frac{1}{2\pi i}\int_{\gamma_2}\frac{f(\zeta)}{(\zeta-a)^{n+1}}d\zeta \\ &\quad + \sum_{n=0}^{\infty}(z-a)^{-(n+1)}\frac{1}{2\pi i}\int_{\gamma_1}\frac{f(\zeta)}{(\zeta-a)^{-n}}d\zeta\,. \end{aligned}$$

Now the second series is easily seen to be

$$\sum_{n=-1}^{-\infty}(z-a)^n \frac{1}{2\pi i}\int_{\gamma_1}\frac{f(\zeta)}{(\zeta-a)^{n+1}}d\zeta\,.$$

This is the same as summing, $\sum_{n=-\infty}^{-1}$. Hence, if for $n \geq 0$ we take

$$a_n = \frac{1}{2\pi i} \int_{\gamma_2} \frac{f(\zeta)}{(\zeta - a)^{n+1}} d\zeta\,,$$

and for $n < 0$,

$$a_n = \frac{1}{2\pi i} \int_{\gamma_1} \frac{f(\zeta)}{(\zeta - a)^{n+1}} d\zeta\,,$$

we get the Laurent expansion.

Since the integrand $\frac{f(\zeta)}{(\zeta-a)^{n+1}}$ is holomorphic in the region bounded by the two circles, Cauchy's theorem for a simply connected region tells us we can replace either of these circles by a fixed circle, γ, between the two bounding circles of the original annulus. Indeed, we can replace these contours by integrals along any positively oriented piecewise smooth simple closed curve , γ in the annulus (which contains a in its interior). Hence we have the following:

Theorem 4.5.1 *Each holomorphic function in the annulus is given by*

$$f(z) = \sum_{n=-\infty}^{\infty} a_n(z-a)^n\,,$$

where for all $n \in \mathbb{Z}$,

$$a_n = \frac{1}{2\pi i} \int_{\gamma} \frac{f(\zeta)}{(\zeta - a)^{n+1}} d\zeta\,.$$

Of course, for $n \geq 0$, a_n is not equal to $\frac{f^{(n)}(a)}{n!}$. In fact, f may not be holomorphic at a at all. Indeed, the following are consequences of the Laurent expansion:

Corollary 4.5.2 (i) *f is holomorphic at a (i.e. a is a removable singularity) if and only if $a_n = 0$ for all $n < 0$. Thus in this case the Laurent expansion is just the Taylor expansion.*

(ii) *f has a pole of order m if and only if $a_m \neq 0$ for some $m < 0$ and $a_n = 0$ for all $n < m$.*

(iii) *f has an essential singularity at a if and only if $a_n \neq 0$ for infinitely many negative n.*

A further consequence is:

Corollary 4.5.3 *Any essential singularity of h at a can be gotten by taking a holomorphic function f defined in a neighborhood of a which is not a polynomial and looking at*

$$f\left(\frac{1}{z-a}\right)+g(z),$$

where g is holomorphic in a neighborhhod of a. To get a pole of order m, do the same, but this time let f be a polynomial of degree m.

4.6 The calculation of residues at an isolated singularity, the residue theorem

Let Ω be a domain, a be a point in it and suppose f is holomorphic on $\Omega - \{a\}$. Consider any positively oriented piecewise smooth simple closed curve γ in Ω with a in its interior. We now define the residue of f at a (cf. Figure 4.7).

Definition 4.6.1 $\text{Res}(f,a) = \frac{1}{2\pi i}\int_\gamma f(z)dz.$

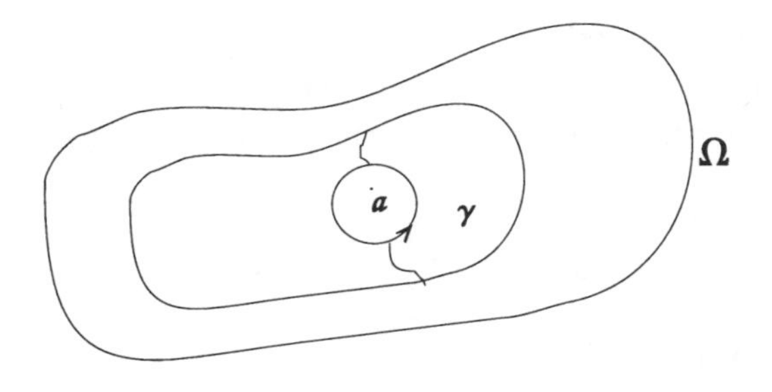

Figure 4.7

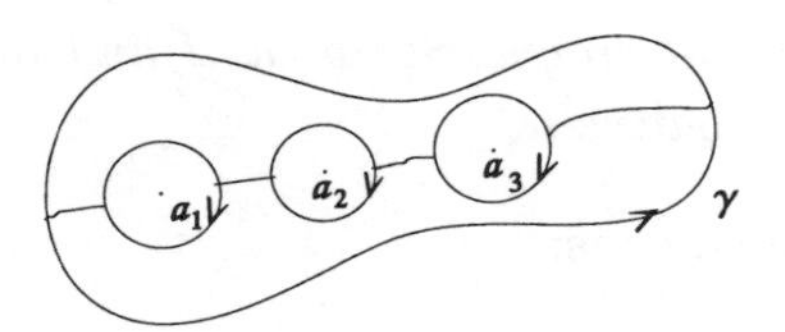

Figure 4.8

Notice that the residue is actually independent of γ and depends only on f and a. This is because, by the pre-residue theorem, we can replace γ in the integral above by a small circle γ_0 centered at a. Thus if γ_* is another curve with the same properties as γ then

$$\frac{1}{2\pi i}\int_\gamma f(z)dz = \frac{1}{2\pi i}\int_{\gamma_0} f(z)dz = \frac{1}{2\pi i}\int_{\gamma_*} f(z)dz\,.$$

The importance of this notion is to be found in the following, called the residue theorem.

Theorem 4.6.2 *Let f be a meromorphic function on a domain Ω and γ be a positively oriented, piecewise smooth, simple closed curve in Ω. Suppose $\{a_1, \ldots, a_n\}$ are the singularities of f inside γ. Then*

$$\int_\gamma f(z)dz = 2\pi i \sum_{j=1}^{n} \operatorname{Res}(f, a_j)\,.$$

In this way we can reduce computations of contour integrals to finding isolated singularities and computing their residues (cf. Figure 4.8). This result follows immediately from the pre-residue theorem (cf. Corollary 3.1.2) and the definition of a residue given above.

Now let a be an isolated singularity of f and consider the Laurent expansion

$$f(z) = \sum_{-\infty}^{+\infty} a_n (z-a)^n\,.$$

Then for γ, as above,

$$\frac{1}{2\pi i}\int_\gamma f(z)dz = \frac{1}{2\pi i}\int_\gamma \sum_{-\infty}^{+\infty} a_n (z-a)^n dz = \frac{1}{2\pi i}\sum_{-\infty}^{+\infty} a_n \int_\gamma (z-a)^n dz\,.$$

If $n \neq -1$ the last integral is zero. When $n = -1$ it is $2\pi i$. It follows that $\text{Res}(f, a) = a_{-1}$. Of course, if a is a removable singularity then by Cauchy's theorem $\text{Res}(f, a) = 0$ (or by Taylor's theorem $a_{-1} = 0$). However, as we have just seen, the residue at a could be zero even if the function were not holomorphic at a.

A more interesting example is $f(z) = \frac{\sin z}{z^2}$. Here $f(z) = \frac{1}{z} - \frac{z}{3!} + \frac{z^3}{5!} \cdots$ and because $\frac{\sin z}{z}$ has a removable singularity, f has a simple pole at $z = 0$ and $a_{-1} = 1$.

We shall be especially interested in the residue at a pole. Suppose a is a pole of f of order m. Then the Laurent expansion of f at a is

$$f(z) = \frac{a_m}{(z-a)^m} + \frac{a_{m-1}}{(z-a)^{m-1}} \cdots + \frac{a_{-1}}{z-a} + \phi(z),$$

where ϕ is holomorphic. Let $g(z) = (z-a)^m f(z)$. Then $g(z)$ is given by the convergent power series $g(z) = a_m + a_{m-1}(z-a) \cdots + a_{-1}(z-a)^{m-1} + \cdots$. Hence g is holomorphic at a and $a_{-1} = \frac{g^{m-1}(a)}{(m-1)!}$. In terms of f, the latter is just

$$\lim_{z \to a} \frac{d^{m-1}}{dz^{m-1}} \left(\frac{(z-a)^m f(z)}{(m-1)!} \right).$$

This leads us to the following:

Proposition 4.6.3

$$\text{Res}(f, a) = \frac{g^{m-1}(a)}{(m-1)!} = \lim_{z \to a} \frac{d^{m-1}}{dz^{m-1}} \left(\frac{(z-a)^m f(z)}{(m-1)!} \right).$$

As examples, we can use this proposition when $f(z) = \frac{z^3+5}{z(z-1)^3}$. Here f has a pole of order 3 at $z = 1$ and $\text{Res}(f, 1) = 6$. Similarly, if $f(z) = \frac{ze^z}{(z-1)^2}$, then $z = 1$ is a pole of order 2 with residue $\text{Res}(f, 1) = 2e$.

In the case of a simple pole the formula just says

$$\text{Res}(f, a) = \lim_{z \to a} (z-a) f(z).$$

However, a more convenient formula for the residue of a simple pole is contained in the next proposition.

Proposition 4.6.4 *Let $f = \frac{\phi(z)}{\psi(z)}$, where ϕ and ψ are holomorphic and a is a first order pole. Then* $\operatorname{Res}(f, a) = \frac{\phi(a)}{\psi'(a)}$.

Proof. Expanding in Taylor series about $z = a$ gives $\phi(z) = a_0 + a_1(z - a) + a_2(z-a)^2 + \cdots$ and $\psi(z) = b_0 + b_1(z-a) + b_2(z-a)^2 + \cdots$. Since $\phi(a) \neq 0$, $\psi(a) = 0$ and $\psi'(a) \neq 0$, we see $a_0 \neq 0$, $b_0 = 0$ and $b_1 \neq 0$. Hence,

$$g(z) = (z-a)f(z) = \frac{a_0(z-a) + a_1(z-a)^2 + a_2(z-a)^2 + \cdots}{b_1(z-a) + b_2(z-a)^2 + \cdots}.$$

Clearly, the limiting value of this as $z \to a$ is $\frac{a_0}{b_1} = \frac{\phi(a)}{\psi'(a)}$. □

So, for example, if $f(z) = \frac{z}{z^2-1}$, then f has simple poles at ± 1, both with residue $\frac{1}{2}$. If $f(z) = \frac{\tan(z)}{z} = \frac{\sin(z)}{z\cos(z)}$, then, as we saw earlier, $z = 0$ is a removable singularity of f and $z = \pm\frac{\pi}{2}, \pm\frac{3\pi}{2}\ldots$, are simple zeros of cos and therefore simple poles of f. For each n, $\operatorname{Res}(f, \pm\frac{(2n-1)\pi}{2}) = \mp\frac{2}{(2n-1)\pi}$.

It follows from $\operatorname{Res}(f, a) = -\operatorname{Res}(f, -a)$ that if γ is a positively oriented circle centered at the origin, not passing through any of these poles, then $\int_\gamma \frac{\tan(z)}{z}dz = 0$.

Finally, using the same ideas, we derive a formula similar to the one just above in the case of a 2nd order pole.

Proposition 4.6.5 *Let $f(z) = \frac{\phi(z)}{\psi(z)}$, where ϕ and ψ are holomorphic and a be a second order pole of f. Then*

$$\operatorname{Res}(f, a) = \frac{(\psi^{(2)}(a)/2!)(\phi^{(1)}(a)/1!) - \phi(a)\psi^{(3)}(a)/3!}{(\psi^{(2)}(a)/2!)^2}.$$

Proof.

$$g(z) = (z-a)^2 f(z) = (z-a)^2\frac{\phi(z)}{\psi(z)} = \frac{a_0 + a_1(z-a) + \cdots}{b_2 + b_3(z-a) + \cdots}.$$

Calculating $\frac{d}{dz}g(z)$ and taking the limit as $z \to a$ yields $\frac{b_2 a_1 - a_0 b_3}{b_2^2}$. □

As an example, if $f(z) = \frac{e^z}{1-\cos(z)}$, then $z = 0$ is a second order pole and $\text{Res}(f, 0) = 2$.

Exercise 4.4 1. *Calculate* $\int_\gamma z^n e^{\frac{1}{z}} dz$, *where* γ *is a circle centered at zero of radius* $r > 0$.

2. *Let* f *and* g *be meromorphic functions defined on the same domain. Show that, for any point* a *in the domain,* $\text{Res}(f + g, a) = \text{Res}(f, a) + \text{Res}(g, a)$.

4.7 Application to the calculation of real integrals

It is an interesting feature of complex analysis that it enables one to calculate certain definite real integrals, especially those useful in Fourier analysis, both Fourier series and Fourier integrals.

We first consider integrals of the form $\int_0^{2\pi} F(\cos\theta, \sin\theta) d\theta$, where F is a rational function of two variables. We shall illustrate the situation with the following typical example: $\int_0^{2\pi} \frac{1}{a+b\cos\theta} d\theta$, where $a > b > 0$.

First observe that $a + b\cos\theta \neq 0$ for any θ since, if it were, we would get $|\cos\theta| = \frac{|a|}{|b|} > 1$, a contradiction. Now since $e^{i\theta} = \cos\theta + i\sin\theta$, we see that $e^{-i\theta} = \cos\theta - i\sin\theta$. Hence, $\cos\theta = \frac{1}{2}(e^{i\theta} + e^{-i\theta})$ and $\sin\theta = \frac{1}{2i}(e^{i\theta} - e^{-i\theta})$.

Let $z = e^{i\theta}$. Then $dz = izd\theta$. If γ denotes the positively oriented unit circle, $\gamma(\theta) = e^{i\theta}$, $0 \leq \theta \leq 2\pi$, then our integral is just

$$\frac{2}{i} \int_\gamma \frac{dz}{bz^2 + 2az + b}.$$

It is easy to see that the roots of this quadratic denominator are both real; one of them lies inside and one outside the unit circle. On the inside, we have $-1 < \frac{-a+\sqrt{a^2-b^2}}{b} = \alpha < 1$. Thus our integrand has a simple pole at α. Using Proposition 4.6.4 its residue is given by

$$\text{Res}(f, \alpha) = \frac{\phi(\alpha)}{\psi'(\alpha)} = \frac{1}{2\sqrt{a^2 - b^2}}.$$

Hence, by Theorem 4.6.2 our integral is

$$\frac{2}{i}2\pi i\frac{1}{2\sqrt{a^2-b^2}}=\frac{2\pi}{\sqrt{a^2-b^2}}\,.$$

Lemma 4.7.1 *For* $0\leq\theta\leq\frac{\pi}{2}$, $\sin\theta\geq\frac{2}{\pi}\theta$.

Proof. As we saw, $f(\theta)=\frac{\sin\theta}{\theta}$ has a removable singularity at 0 and, therefore, is analytic in $\mathbb{R}$. Let us consider the minimum value of f on $[0,\frac{\pi}{2}]$. If this occured on the open interval, then it must occur at a critical point and therefore at a point where $\tan(\theta)=\theta$. But as we saw in Section 4.5 there is no such θ since the only root of this equation on $(-\frac{\pi}{2},\frac{\pi}{2})$ is $\theta=0$. This means the minimum value occurs at one of the endpoints where $f(0)=1$ and $f(\frac{\pi}{2})=\frac{2}{\pi}<1$. $\square$

Proposition 4.7.2 $\int_0^\infty\frac{\sin x}{x}dx=\frac{\pi}{2}$.

Proof. First observe that since $\sin x$ and x are both odd functions our integrand is even. Hence $\int_{-\infty}^\infty\frac{\sin x}{x}dx=2\int_0^\infty\frac{\sin x}{x}dx$. Moreover, as we saw just above, $\frac{\sin x}{x}$ is analytic on the $\mathbb{R}$-axis. Using Euler's relation as in the previous integral, we see that $\frac{\sin x}{x}=\frac{1}{2i}\frac{e^{ix}-e^{-ix}}{x}$. Therefore,

$$\int_0^\infty\frac{\sin x}{x}dx=\frac{1}{2i}\int_0^\infty\frac{e^{ix}-e^{-ix}}{x}dx=\lim_{r\to\infty,\epsilon\to0}\frac{1}{2i}\int_\epsilon^r\frac{e^{ix}-e^{-ix}}{x}dx\,.$$

Now, considering the map $x\mapsto -x$ with derivative -1, we get

$$\int_\epsilon^r\frac{e^{ix}-e^{-ix}}{x}dx=\int_\epsilon^r\frac{e^{ix}}{x}dx+\int_{-r}^{-\epsilon}\frac{e^{-ix}}{x}dx\,,$$

so that

$$\int_0^\infty\frac{\sin x}{x}dx=\frac{1}{2i}\lim_{r\to\infty,\epsilon\to0}\int_\epsilon^r\frac{e^{ix}}{x}dx+\int_{-r}^{-\epsilon}\frac{e^{-ix}}{x}dx\,.$$

Therefore, we consider the function $\frac{e^{-iz}}{z}$ and its contour integral over γ, the positively oriented, piecewise smooth, simple closed curve consisting of the semicircles γ_r, γ_ϵ and the intervals, $\gamma_+=[\epsilon,r]$ and

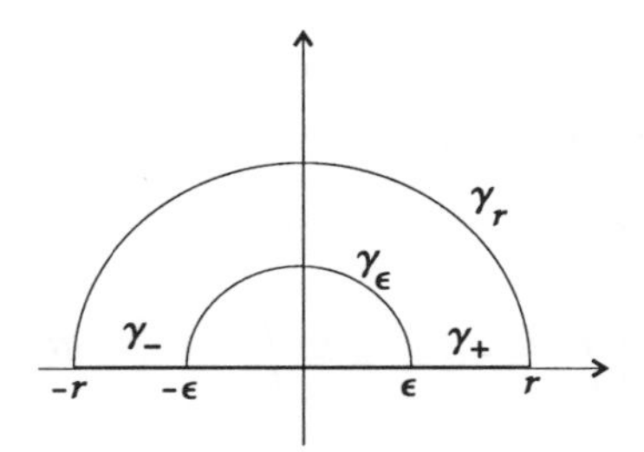

Figure 4.9

$\gamma_- = [-r, -\epsilon]$ (cf. Figure 4.9). In the interior of this closed curve $\frac{e^{-iz}}{z}$ is holomorphic so, by Cauchy's theorem, $\int_\gamma = 0$. On the other hand, $\int_\gamma = \int_{\gamma_r} + \int_{\gamma_\epsilon} + \int_{\gamma_+} + \int_{\gamma_-}$. Now $\int_{\gamma_+} + \int_{\gamma_-} \to \int_{-\infty}^{\infty} \frac{\sin x}{x}$, as $r \to \infty$ and $\epsilon \to 0$. We will show:

1. $\lim_{r\to\infty} \int_{\gamma_r} = 0$ and

2. $\lim_{\epsilon\to 0} \int_{\gamma_\epsilon} = -i\pi$.

Then,

$$2i \int_0^\infty \frac{\sin x}{x} = \lim_{r\to\infty,\epsilon\to 0} \left(\int_{\gamma_+} + \int_{\gamma_-} \right) = \int_\gamma - \int_{\gamma_r} - \int_{\gamma_\epsilon} = \pi i\,.$$

So $\int_0^\infty \frac{\sin x}{x} = \frac{\pi}{2}$.

Turning to the proof of statement 1, here we are interested in estimating $|\int_{\gamma_r} f(z)e^{iz}dz|$, particularly where $f(z) = \frac{1}{z}$, but perhaps also for other functions. Now the usual estimate gives $|\int| \le \pi r M(r)$, where $M(r)$ is the maximum value of $|f(z)|$ on the semicircular arc in the upper half plane of radius r. In our case, $M(r) = \frac{1}{r}$ and so this estimate is *not nearly good enough* to prove statement 1. We change to polar coordinates, $z = re^{i\theta}$, where $0 \le \theta \le \pi$. Then $\int_{\gamma_r} \frac{e^{iz}}{z} dz = \int_0^\pi e^{ir\cos\theta - r\sin\theta} i d\theta$. Hence, $|\int_{\gamma_r} \frac{e^{iz}}{z} dz| \le \int_0^\pi e^{-r\sin\theta} d\theta$, and since $\sin\theta = \sin(\pi - \theta)$, it follows that

$$\int_0^\pi e^{-r\sin\theta} d\theta = 2\int_0^{\frac{\pi}{2}} e^{-r\sin\theta} d\theta\,.$$

Hence, $|\int_{\gamma_r} \frac{e^{iz}}{z} dz| \leq 2\int_0^{\frac{\pi}{2}} e^{-r\sin\theta} d\theta$. But on $[0, \frac{\pi}{2}]$ we know, by the previous lemma, that $\sin\theta \geq \theta\frac{2}{\pi}$. Therefore, $-r\sin\theta \leq -\frac{2r\theta}{\pi}$ and since exp is monotone increasing on the real axis $e^{-r\sin\theta} \leq e^{-\frac{2r\theta}{\pi}}$. So

$$\left|\int_{\gamma_r} \frac{e^{iz}}{z} dz\right| \leq 2\int_0^{\frac{\pi}{2}} e^{-\frac{2r\theta}{\pi}} d\theta\,.$$

The latter integral can be evaluated exactly. Its value is $\pi\frac{1-e^{-r}}{r}$. By l'Hôpital's rule its limit as $r \to \infty$ is zero, proving statement 1.

Before turning to statement 2, it is worthwhile considering a somewhat more general question, namely, what can one say about estimating $|\int_{\gamma_r} f(z)e^{iz}dz|$? Clearly, by the argument just given, this is $\leq \pi M(r)(1-e^{-r})$, where $M(r)$ is the maximum value of $|f(z)|$ on γ_r. This gives us a way of being sure that in certain other cases $|\int_{\gamma_r} f(z)e^{iz}dz|$ tends to zero as $r \to \infty$.

Proof of statement 2. Expand e^{iz} in a power series about 0, $e^{iz} = 1 + iz + z\eta(z)$, where $\eta(z) \to 0$ as $z \to 0$. Hence $\frac{e^{iz}}{z} = \frac{1}{z} + i + \eta(z)$. Integrating and using linearity we get

$$\int_{\gamma_\epsilon} \frac{e^{iz}}{z} dz = \int_{\gamma_\epsilon} \frac{1}{z} dz + \int_{\gamma_\epsilon} i dz + \int_{\gamma_\epsilon} \eta(z) dz\,.$$

Now, $\int_{\gamma_\epsilon} idz = i(\epsilon - (-\epsilon)) = 2i\epsilon$ which tends to zero. Also, $\int_{\gamma_\epsilon} \eta(z)dz \leq L(\gamma_\epsilon)\max|\eta(z)| = \pi\epsilon\max|\eta(z)|$, which certainly tends to zero. Hence the limiting value of $\int_{\gamma_\epsilon} \frac{e^{iz}}{z} dz$ is

$$\int_\pi^0 \frac{\epsilon e^{i\theta}}{\epsilon e^{i\theta}} i d\theta = \int_\pi^0 i d\theta = -\pi i\,. \qquad \square$$

We now use these ideas to calculate some other integrals, for example, $\int_{-\infty}^{\infty} \frac{x\sin x}{x^2+4} dx$.

Consider $\int_{-\infty}^{\infty} \frac{xe^{ix}}{x^2+4} dx$ and the function $f(z) = \frac{z}{z^2+4}$ in the upper half plane. Here $f(z) \to 0$ as $z \to \infty$. In the upper half plane f has singularities only at $z = 2i$, which is a simple pole. As we saw, $|\int_{\gamma_r} f(z)e^{iz}dz| \leq \pi M(r)(1-e^{-r})$, where $M(r)$ is the maximum value of

$|f(z)|$ on γ_r. Here for r sufficiently large $M(r)$ is of the order of $\frac{1}{r}$ and so $|\int_{\gamma_r} f(z)e^{iz}dz|$ tends to zero as $r \to \infty$. Consider the simple closed curve gotten from γ_r together with the interval $[-r, r]$. Applying the residue theorem we get

$$\int_{\gamma_r} f(z)e^{iz}dz + \int_{-r}^{r} \frac{xe^{ix}}{x^2+4} = 2\pi i \operatorname{Res}\left(\frac{ze^{iz}}{z^2+4}, 2i\right).$$

A direct calculation of the residue yields $\frac{1}{2}e^{-2}$. Using Euler's relation and going back to real integrals we get

$$\int_{-r}^{r} \frac{xe^{ix}}{x^2+4} = \int_{-r}^{r} \frac{x\cos x}{x^2+4} + i\int_{-r}^{r} \frac{x\sin x}{x^2+4}.$$

Since by symmetry the term involving cos is zero, it follows that $\int_{-\infty}^{\infty} \frac{x\sin x}{x^2+4}dx = \pi e^{-2}$.

More generally, suppose we are interested in calculating $\int_{-\infty}^{\infty} \frac{e^{i\lambda x}}{p(x)}dx$, where p is a polynomial of degree ≥ 2 having no real roots and λ is non-negative. Hence, $\frac{e^{i\lambda x}}{p(x)}$ is integrable over $\mathbb{R}$. Now $f(z) = \frac{e^{i\lambda z}}{p(z)}$ is meromorphic on the upper half plane and has no singularities on $\mathbb{R}$. As above, let $M(r) = \max\{|f(z)| : |z| = r, \Im z \geq 0\}$. If $z = x + iy$, then $|e^{i\lambda(x+iy)}| = e^{-\lambda y}$ so that $|f(z)| \leq \frac{e^{-\lambda y}}{|p(z)|}$. On the upper half plane $e^{-\lambda y} \leq 1$ since $y \geq 0$ and $\lambda > 0$. Because $\frac{1}{|p(z)|} \to 0$, we see that $M(r) \to 0$ and, therefore, $\pi M(r)(1 - e^{-r})$ also tends to zero as $r \to \infty$. It follows that

$$\int_{-\infty}^{\infty} \frac{e^{i\lambda x}}{p(x)}dx = 2\pi i \Sigma \operatorname{Res}(f, p),$$

where we sum over all the poles p of f in the upper half plane. So, for example, $\int_{-\infty}^{\infty} \frac{e^{i\lambda x}}{1+x^2}dx = 2\pi i \frac{e^{i\lambda i}}{2i} = \pi e^{-\lambda}$.

Of course, if $\lambda = 0$, then this works even better. As above, $M(r) \to 0$ as $r \to \infty$. So, for example, $\int_{-\infty}^{\infty} \frac{dx}{1+x^4} = 2\pi i(\operatorname{Res}(f, \frac{-1+i}{\sqrt{2}}) + \operatorname{Res}(f, \frac{1+i}{\sqrt{2}}))$. Computing these residues (of simple poles) by the method above yields $\int_{-\infty}^{\infty} \frac{dx}{1+x^4} = \frac{\pi}{\sqrt{2}}$.

Finally, here is a variant of this scheme. Let $f(x) = \frac{p(x)}{q(x)}$, where p and q are polynomials with real coefficients such that q has no zeros on the real axis and $\deg q \geq \deg p + 2$. Then $\int_{-\infty}^{\infty} f(x)dx$ exists because, for $|z|$ large enough, $|f(z)| \leq \frac{c}{|z|^2}$, where c is a constant. Hence the tail is an L^2 function. Also $|\int_{\gamma_r} f(z)dz| \to 0$ as $r \to \infty$. So $\int_{-\infty}^{\infty} f(x)dx = 2\pi i \Sigma \operatorname{Res}(f, p)$, where again we sum over all the poles of f in the upper half plane. As an example, we get $\int_{-\infty}^{\infty} \frac{x^2}{x^6+1}dx = \frac{2\pi}{3}$.

4.8 A more general removable singularities theorem and the Schwarz reflection principle

We begin this section with a removable singularities theorem generalizing Riemann's theorem where the removable point singularity is replaced by an interval of possible singularities. In the earlier situation boundedness in a neighborhood of the singularity was the necessary hypothesis and was equivalent to continuity at that point.

Theorem 4.8.1 *Let Ω be a domain, I be a finite closed interval $I \subseteq \Omega$, and $f : \Omega \to \mathbb{C}$ be a function which is holomorphic on $\Omega - I$ and continuous on I. Then f is actually holomorphic on Ω.*

Proof. Let $a \in I$. We will show f is holomorphic at a. If a happens to be an endpoint of I, just extend I slightly while remaining in Ω. Thus we can assume a is an interior point of I. Since being holomorphic is a local property, we can restrict f to a smaller subinterval with a its center and replace Ω by an open rectangle $\mathcal{R}$, with this subinterval a horizontal line dividing it in half. Thus a is the center of $\mathcal{R}$ and the sides of $\mathcal{R}$ are parallel or perpendicular to I. Let γ be the boundary of $\mathcal{R}$ with positive orientation. This is a piecewise smooth, simple closed curve (cf. Figure 4.10). Now f is continuous on Ω; for $z \in \mathcal{R}$, but off the trajectory of γ the function $\zeta \mapsto \zeta - z$ is continuous on γ and never zero. Hence, $\zeta \mapsto \frac{f(\zeta)}{\zeta - z}$ is continuous on γ and therefore $\frac{1}{2\pi i} \int_\gamma \frac{f(\zeta)}{\zeta - z} = g(z)$ exists for each z in the interior of $\mathcal{R}$ (which we will now call $\mathcal{R}$). Moreover, as we proved in Proposition 3.2.6, g is holomorphic on $\mathcal{R}$. We will show that on $\mathcal{R}$, $f \equiv g$. Then f is holomorphic on $\mathcal{R}$ and, in particular, at a.

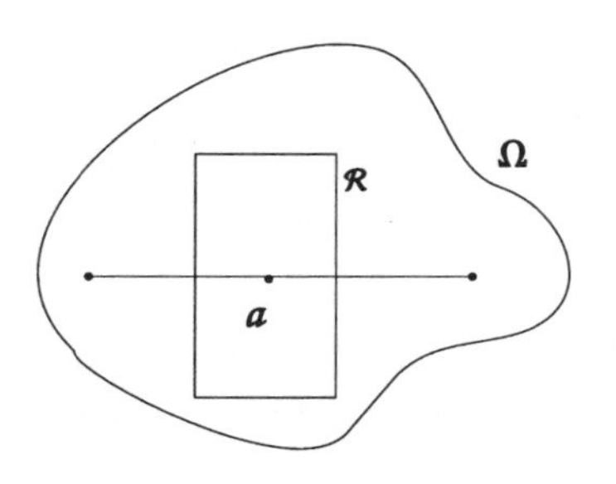

Figure 4.10

We shall refer to $\mathcal{R} \cap I$ simply as I. Let $z \in \mathcal{R} - I$. Then $\int_I \frac{f(\zeta)}{\zeta - z}$ exists, going in either direction. Let α and β be the positively oriented contours gotten by going around the upper and lower halves of $\mathcal{R}$. Then $\int_\alpha \frac{f(\zeta)}{\zeta - z}$ and $\int_\beta \frac{f(\zeta)}{\zeta - z}$ also exist and cancel along I so that $g(z) = \frac{1}{2\pi i} \int_\alpha \frac{f(\zeta)}{\zeta - z} + \frac{1}{2\pi i} \int_\beta \frac{f(\zeta)}{\zeta - z}$. Now choose two sequences of contours $\alpha_n \to \alpha$ and $\beta_n \to \beta$ whose sides coincide with those of α and β, respectively, but which, on the side nearest to I, are a distance $\frac{1}{n}$ from I. Evidently,

$$\frac{1}{2\pi i} \int_{\alpha_n} \frac{f(\zeta)}{\zeta - z} \to \frac{1}{2\pi i} \int_\alpha \frac{f(\zeta)}{\zeta - z},$$

and

$$\frac{1}{2\pi i} \int_{\beta_n} \frac{f(\zeta)}{\zeta - z} \to \frac{1}{2\pi i} \int_\beta \frac{f(\zeta)}{\zeta - z},$$

since in each case the difference approaches $\frac{1}{2\pi i} \int_I \frac{f(\zeta)}{\zeta - z} + \frac{1}{2\pi i} \int_{-I} \frac{f(\zeta)}{\zeta - z} = 0$. Thus,

$$g(z) = \lim_{n \to \infty} \left(\frac{1}{2\pi i} \int_{\alpha_n} \frac{f(\zeta)}{\zeta - z} + \frac{1}{2\pi i} \int_{\beta_n} \frac{f(\zeta)}{\zeta - z} \right).$$

If z is in the top half of $\mathcal{R}$ since f is holomorphic on $\mathcal{R} - I$, we see that for n sufficiently large, $f(z) = \frac{1}{2\pi i} \int_{\alpha_n} \frac{f(\zeta)}{\zeta - z}$, while $\frac{1}{2\pi i} \int_{\beta_n} \frac{f(\zeta)}{\zeta - z} = 0$ for all n. Hence, for n sufficiently large, $f(z) = \frac{1}{2\pi i} \int_{\alpha_n} \frac{f(\zeta)}{\zeta - z} + \frac{1}{2\pi i} \int_{\beta_n} \frac{f(\zeta)}{\zeta - z}$. This means that f and g coincide on the top half of $\mathcal{R}$. Similarly, they coincide on the bottom half of $\mathcal{R}$. Thus, g and f coincide on $\mathcal{R} - I$.

Since these functions are both continuous on $\mathcal{R}$, and $\mathcal{R} - I$ is dense in $\mathcal{R}$, they coincide everywhere on $\mathcal{R}$. □

Before turning to the Schwarz reflection principle we need the following lemma.

Lemma 4.8.2 *Let Ω be a domain and J the conjugation map. Then $J(\Omega)$ is a domain in $\mathbb{C}$ and, if $f : \Omega \to \mathbb{C}$ is a holomorphic function and g is defined on $J(\Omega)$ by $g(z) = \overline{f(\bar{z})}$, $z \in J(\Omega)$, then g is holomorphic on $J(\Omega)$.*

Proof. It is clear that $J(\Omega)$ is a domain. Let z and $a \in J(\Omega)$. Then $\bar{z}$ and $\bar{a} \in \Omega$. Since f is holomorphic on Ω, we know $f(\bar{z}) = \sum_{n=0}^{\infty} a_n(\bar{z} - \bar{a})^n$. Hence, $\overline{f(\bar{z})} = g(z) = \sum_{n=0}^{\infty} \overline{a_n}(z - a)^n$ and so g is holomorphic at every point of $J(\Omega)$. □

Theorem 4.8.3 *Let Ω be a domain in the upper half plane whose boundary includes an interval $(a, b) = I$ on the real axis. Let $f : \Omega \to \mathbb{C}$ be a holomorphic function which is continuous on I. Suppose $f(x + 0i)$ is real for all $x \in I$ and define*

$$F(z) = f(z), z \in \Omega \cup I$$

and

$$F(z) = \overline{f(\bar{z})}, z \in J(\Omega).$$

Then F is a holomorphic function on $\Omega \cup I \cup J(\Omega)$ which extends f on Ω.

Proof. Clearly, $(\Omega \cup I) \cap J(\Omega)$ is empty so F is well defined. Evidently, $F|(\Omega \cup I) = f$. By Lemma 4.8.2, $z \mapsto \overline{f(\bar{z})}$ is holomorphic on $J(\Omega)$ and since Ω and $J(\Omega)$ are disjoint, F is holomorphic on $\Omega \cup J(\Omega)$. By our removable singularities theorem just above, to complete the proof we need only see that F is continuous on I. Let $x \in I$ and $z_n \to x$, where $z_n \in \Omega \cup I \cup J(\Omega)$. If all but finitely many z_n are in $\Omega \cup I$, there is no problem since here F agrees with f, which is continuous so $F(z_n) \to F(x)$. Therefore, we may assume infinitely many $z_n \in J(\Omega)$ and by choosing subsequences and knowing we are fine on Ω, we may

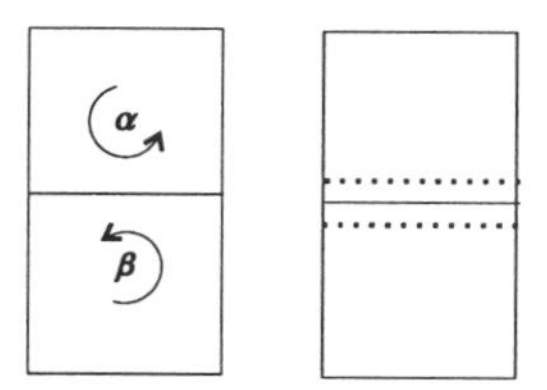

Figure 4.11

actually assume they all lie in $J(\Omega)$. Then $F(z_n) = \overline{f(\overline{z_n})}$, where $\overline{z_n} \in \Omega$. Now $\overline{z_n} \to \bar{x} = x$ so, by continuity, $f(\overline{z_n}) \to f(x)$. Hence, taking conjugates and using the fact that the boundary values of f are real, $F(z_n) = \overline{f(\overline{z_n})} \to \overline{f(x)} = f(x) = F(x)$ (cf. Figure 4.11). □

Remark The removable singularities theorem actually works for any smooth arc γ, not just an interval. This can be proved along the lines of the classical Theorem 4.8.1 by making use of the inequality involving C^1 convergence used to prove Cauchy theorem for a simply connected domain. The Schwarz reflection principle itself generalizes as follows: Let Ω_1 and Ω_2 be disjoint domains whose closures, $\bar{\Omega}_1$ and $\bar{\Omega}_2$ have the property that $\bar{\Omega}_1 \cap \bar{\Omega}_2$ is a simple smooth curve, γ. Suppose for $j = 1, 2$, f_j is holomorphic on Ω_j and continuous on the trajectory of γ. If for each x on the trajectory of γ, we have $\lim_{z \in \Omega_1 \to x} f_1(z) = \lim_{z \in \Omega_2 \to x} f_2(z)$, then there exists a holomorphic function F on $\Omega_1 \cup \gamma \cup \Omega_2$ such that, for $j = 1, 2$, $F|\Omega_j = f_j$. This is particularly useful in the case of Schwarz reflection in the unit circle, $z \mapsto \frac{1}{\bar{z}} = \frac{z}{|z|}$ and will be used in the last chapter in dealing with the unitary group $\mathrm{U}(p, q)$ of a non-degenerate, but indefinite Hermitian form.

Exercise 4.5 *Suppose $f(z) = \sum_{n=0}^{\infty} a_n z^n$ is convergent for $|z| < r$ and $f(x)$ is real for $-r < x < r$. Show all a_n are real and, for all $|z| < r$, $f(\bar{z}) = \overline{f(z)}$.*

Chapter 5

Conformal Mappings

The reader will notice in this chapter that group theory plays a role.

5.1 Linear fractional transformations, equivalence of the unit disk and the upper half plane

First we gather some useful facts. From a 2×2 matrix with entries from $\mathbb{C}$,

$$\begin{pmatrix} a & b \\ c & d \end{pmatrix}$$

we form the fractional linear transformation $g(z) = \frac{az+b}{cz+d}$. When there is no danger of confusion, we sometimes also refer to the matrix above as g. Notice however that multiplying the matrix by $\lambda \neq 0$ changes the matrix, but leaves the fractional linear transformation unchanged. For $c \neq 0$, this is a holomorphic map on all of $\mathbb{C}$ except for $z = -\frac{d}{c}$. If it happens that $c = 0$ and $d \neq 0$, then it is holomorphic everywhere. If both c and d are zero, this does not define a mapping at all.

As is customary we denote the group of 2×2 invertible complex matrices by $\mathrm{GL}(2, \mathbb{C}) = G$ and the projective group gotten by dividing by its center (the scalar matrices) by $\mathrm{PGL}(2, \mathbb{C})$.

Exercise 5.1 *Verify that g is an invertible fractional linear transformation if and only if* $\det(g) \neq 0$. *Because we can multiply by $\lambda \neq 0$ and leave the fractional linear transformation unchanged, if* $\det g \neq 0$, *we can normalize things by taking* $\det g = 1$. *Thus, the group that is really acting on $\mathbb{C}$ by linear fractional transformations is not* $\mathrm{GL}(2,\mathbb{C}) = G$, *but rather* $\mathrm{PGL}(2,\mathbb{C})$. *Prove the invertible fractional linear transformations (called Möbius transformations) form a group and this is isomorphic with* $\mathrm{PGL}(2,\mathbb{C})$.

Now let us consider fractional linear transformations $g(z) = \frac{az+b}{cz+d}$, where $\det(g) = 1$. Then one checks easily that $g'(z) = \frac{1}{(cz+d)^2}$. Another useful and easily verifiable fact is, if g is a real matrix, i.e. is in $\mathrm{SL}(2,\mathbb{R})$, then $\Im g(z) = \frac{\Im z}{|cz+d|^2}$. We now define the Cayley transform $c(z) = \frac{z-i}{z+i}$. Since the determinant is $2i$, this fractional linear transformation is invertible. A direct calculation shows the inverse Cayley transform is $c^{-1}(w) = i\frac{1+w}{1-w}$.

Proposition 5.1.1 *The Cayley transform is a biholomorphic equivalence between the upper half plane H^+ and the unit disk D.*

Proof. Since the only singularity of c occurs at $z = -i$, c is holomorphic on the upper half plane. That $\det c \neq 0$ shows it is invertible. If we can prove it maps onto D, then since the inverse is automatically holomorphic, we would be done. Now $|z - i| < |z + i|$. To see this, just draw the picture or square both terms. The latter reduces the question to checking that $-2\Im z < 2\Im z$, which is true since $\Im z > 0$. This proves $\frac{|z-i|}{|z+i|} < 1$ and, therefore, $c(H^+) \subseteq D$. Now let $w \in D$. Then $w \neq 1$. Let $z = i\frac{1+w}{1-w}$. Then, as we saw, $z = c^{-1}(w)$ so $c(z) = w$. □

5.2 Automorphism groups of the disk, upper half plane and entire plane

For a domain Ω we denote by $\mathrm{Aut}(\Omega)$ the set of all holomorphic automorphisms, i.e. bijective holomorphic maps (which automatically have holomorphic inverses by Corollary 1.4.7). These clearly form a group.

Actually, they form a Lie group. We will now calculate these groups in a few important cases.

First we deal with the entire complex plane.

Proposition 5.2.1 $\mathrm{Aut}(\mathbb{C})$ *is the complex* $az+b$ *group.*

Proof. Of course, functions of the form $z \mapsto az+b$ are entire. If $a \neq 0$, they are invertible since we can solve for z. Thus $z = \frac{f(z)-b}{a}$. So the inverse is $g(z) = \frac{z-b}{a}$, which is also holomorphic. Conversely, suppose $f(z) = \sum_{n=0}^{\infty} a_n z^n$ is an entire function which has an entire inverse. Then

$$f\left(\frac{1}{z}\right) = \sum_{n=0}^{\infty} a_n z^{-n}, z \in \mathbb{C} - (0).$$

Here 0 is an isolated singularity of $f(\frac{1}{z})$. As we remarked earlier, if f is not a polynomial this singularity is essential. But then, by the Casorati-Weierstrass Theorem (Theorem 4.1.6), given any $w \in \mathbb{C}$ there is a sequence $z_n \to 0$ so that $f(\frac{1}{z_n}) \to w$. Let $g = f^{-1}$ and apply g getting $gf(\frac{1}{z_n}) = \frac{1}{z_n} \to g(w)$, since g is continuous. Therefore, $|\frac{1}{z_n}| \to |g(w)|$. But since $z_n \to 0$, we know $|\frac{1}{z_n}| \to \infty$. This contradiction proves f is a polynomial. Therefore, so is f'. By the fundamental theorem of algebra, f' must have a zero or be constant. Since f is invertible, f' has no zeros. Therefore, f' is constant and, by Corollary 1.5.4, f is linear, i.e. $f(z) = az+b$. Since f' has no zeros, $a \neq 0$. □

We now turn to the unit disk. The second statement of the proposition below is essentially the Iwasawa decomposition of $\mathrm{Aut}(D) = \mathrm{PSU}(1,1)$.

Proposition 5.2.2 *If* D *is the unit disk, then* $\mathrm{Aut}(D) = \{T_{a,b} : T_{a,b}(z) = \frac{az+b}{\bar{b}z+\bar{a}} : |a| > |b|\}$. *Moreover, any automorphism of* D *is uniquely a product of a rotation and a* T *where* $a = 1$ *(called translations). This latter set is a subgroup and acts transitively on* D.

Proof. First we show any $T = T_{a,b}$ given above acts holomorphically on the unit disk. Now T is a holomorphic map everywhere except at $z = -\frac{\bar{a}}{\bar{b}}$. But for such a z, $|z| = \frac{|a|}{|b|} > 1$. Thus each such T is holomorphic on D. Next suppose $z \in D$. To see that $|T(z)| < 1$ is equivalent to $|az + b|^2 < |\bar{b}z + \bar{a}|^2$. Expand the inequality getting $|a|^2|z|^2 + |b|^2 < |b|^2|z|^2 + |a|^2$; that is, $(|a|^2 - |b|^2)|z|^2 < (|a|^2 - |b|^2)$. But since $|a|^2 - |b|^2 > 0$ and $|z|^2 < 1$, the inequality is true. Thus, these fractional linear transformations are all in $\text{Aut}(D)$.

Next observe that $\text{Aut}(D)$ acts transitively on D. Consider $T_b(z) = \frac{z+b}{\bar{b}z+1}$, where $|b| < 1$, that is $b \in D$. Since $T_b(0) = b$, we see that $\{T_b : b \in D\}$ already acts transitively on D. This set is actually a subgroup, S. In fact, $T_c \circ T_b = T_{\frac{b+c}{1+c\bar{b}}}$. In particular, T_b and T_{-b} are inverses of one another.

Now let $f \in \text{Aut}(D)$. Using transitivity, compose f with the automorphism T_b for some b to get an automorphism, $g = T_b \circ f$ which leaves 0 fixed. By Schwarz' lemma (Proposition 3.7.6) $|g(z)| \leq |z|$ on D. Let $h = g^{-1}$. Then also $h \in \text{Aut}(D)$ and, since $g(0) = 0$, we know $h(0) = 0$. Therefore, also by Schwarz' lemma, $|h(w)| \leq |w|$ for all $w \in D$. Hence, for all $z \in D$, $|z| = |h(g(z))| \leq |g(z)| \leq |z|$. This means that $|h(w)| = |w|$ for all $w \in D$. Again, by Schwarz' lemma, this forces h to be a rotation. Hence, so is its inverse g. Thus we see f is a product of a rotation with an element of S and in particular that f is one of our T's. Therefore, the set of these T is all of $\text{Aut}(D)$. In particular, they form a group. $\square$

It is usually a good idea when there is a transitive action such as $\text{Aut}(D)$ acting on D to see what the stability group is. Here, an easy calculation tells us if T stabilizes 0 it must be a rotation. Thus $\text{Stab}_{\text{Aut}(D)}(0) = K$, the compact group of rotations and, therefore, $\text{Aut}(D)/K = D$. What is $K \cap S$? If $T_b \in S$ then $T_b(0) = b$. But if T_b is also a rotation, then $T_b(0) = 0$. Therefore, $b = 0$ and $T_b = I$. Hence $K \cap S = (I)$. Since, as we know, $\text{Aut}(D) = KS$, this means that every automorphism of D is *uniquely* a product of a rotation and an element of S.

Exercise 5.2 *Show that if f and $g \in Aut(D)$, $f(0) = g(0)$, and $\arg f'(0) = \arg g'(0)$, then $f = g$.*

As it is sometime useful to change models, we now identify $\text{Aut}(H^+)$ with $\text{PSL}(2, \mathbb{R})$.

Corollary 5.2.3 *If H^+ is the upper half plane, then*

$$\mathrm{Aut}(H^+) = \left\{ g(z) = \frac{az+b}{cz+d} : a, b, c, d \in \mathbb{R}, \det g = 1 \right\}.$$

Proof. Since the Cayley transform and its inverse are fractional linear transformations and, as we just saw, so are the elements of $\mathrm{Aut}(D)$, it follows that the same is true of $\mathrm{Aut}(H^+)$. We know, by Proposition 5.1.1, that the Cayley transform intertwines the actions of $\mathrm{Aut}(D)$ on D and $\mathrm{Aut}(H^+)$ on H^+. For $g \in \mathrm{Aut}(H^+)$, let

$$g = \begin{pmatrix} \alpha & \beta \\ \gamma & \delta \end{pmatrix}$$

be its matrix and

$$\begin{pmatrix} a & b \\ \bar{b} & \bar{a} \end{pmatrix}$$

be the matrix of the corresponding automorphism of D. Then, by the above, we get a matrix equation involving these, intertwined by the matrix of the Cayley transform,

$$\begin{pmatrix} 1 & -i \\ 1 & i \end{pmatrix}.$$

This gives four equations, one for each matrix coordinate:

1. $\alpha - i\gamma = a + b$;
2. $\alpha + i\gamma = \overline{(a+b)}$;
3. $\beta - i\delta = -i(a-b)$;
4. $\beta + i\delta = \overline{i(a-b)}$.

We leave it to the reader to check that g must be real. □

As an indication of the significance of H^+ and its automorphism group (as well as D and its automorphism group) to differential geometry, we now define the hyperbolic metric on H^+ and show that $\mathrm{Aut}(H^+)$

acts isometrically on the upper half plane. Actually, H^+ is the unique simply connected, two-dimensional Riemannian manifold of constant negative curvature -1 (called the hyperbolic plane) and $G = \mathrm{PSL}(2,\mathbb{R})$ is its connected isometry group.

We define $ds = \frac{ds_{Euc}}{y}$ or, alternatively, $ds^2 = \frac{dx^2+dy^2}{y^2}$. Thus $ds^2 = \frac{dzd\bar{z}}{\Im^2 z}$. Now G acts on H^+ by $g(z) = \frac{az+b}{cz+d}$, $z \in H^+$, $g \in G$. Since the determinant is 1, $g'(z) = \frac{1}{(cz+d)^2}$. Therefore, $dg(z) = \frac{dz}{(cz+d)^2}$ and, because g is real, also $\overline{dg(z)} = \frac{d\bar{z}}{[\overline{(cz+d)}]^2}$. But then, $dg(z)\overline{dg(z)} = \frac{dzd\bar{z}}{|cz+d|^4}$. Now recalling that $\Im g(z) = \frac{\Im z}{|cz+d|^2}$, it follows that

$$\frac{dg(z)\overline{dg(z)}}{\Im^2 g(z)} = \frac{dzd\bar{z}}{\Im^2 g(z)|cz+d|^4} = \frac{dzd\bar{z}}{\frac{\Im^2(z)|cz+d|^4}{|cz+d|^4}} = \frac{dzd\bar{z}}{\Im^2(z)}.$$

This proves the following:

Corollary 5.2.4 *Every $g \in G$ is an isometry at each point z of H^+.*

Exercise 5.3 *Show $dA = \frac{dxdy}{y^2}$ is an area on H^+ which is* $\mathrm{SL}(2,\mathbb{R})$ *invariant. This can be done using the fact that* $\mathrm{SL}(2,\mathbb{R})$ *preserves the metric, or directly by means of the change of variables formula for double integrals.*

Definition 5.2.5 A *fundamental domain* for a discrete subgroup Γ of $\mathrm{SL}(2,\mathbb{R})$ is a domain Ω in H^+ with the property that (aside from boundary points of Ω) each point of H^+ lies on a unique Γ transform of Ω.

Actually, all we shall need here is the fact that given a point $z \in H^+$ there is always a $g \in \Gamma$ such that $gz \in \Omega$. Such a domain is provided in [3, p. 12].

Exercise 5.4 *Notice that the domain above is not compact since its unbounded. However, it does have finite area. Prove this by estimating the area by that of the larger domain gotten by drawing a horizontal line through the two cusps (cf. Figure 5.1).*

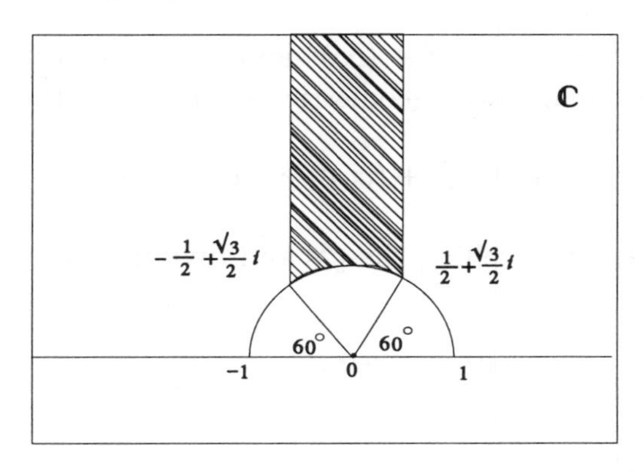

Figure 5.1

We now apply these ideas to prove a result on Diophantine approximation.

Definition 5.2.6 If there is a constant $k > 0$ with the property that for each irrational number α there is a rational number $\frac{p}{q}$ such that

$$|\alpha - \frac{p}{q}| < \frac{k}{q^2},$$

then we say that we have a *diophantine approximation* of α by rationals.

We now use the action on $\mathrm{SL}(2,\mathbb{R})$ on H^+ to produce Diophantine approximations of irrational numbers by means of an idea due to Hürwitz. Let α be an irrational number, which we may clearly assume to be between $-\frac{1}{2}$ and $\frac{1}{2}$, and let $z = \alpha + iy \in H^+$, where $y > 0$ is chosen so that z is not in Ω, the fundamental domain of the modular group, $\mathrm{SL}(2,\mathbb{Z})$ (Figure 5.1). However, because Ω is a fundamental domain, $g(z) \in \Omega$ for some $g \in \mathrm{SL}(2,\mathbb{Z})$. Therefore $\Im g(z) \geq \frac{\sqrt{3}}{2}$. Since $\Im g(z) = \frac{\Im z}{|cz+d|^2}$, c can not be zero. For, if it were, then since $g \in \mathrm{SL}(2,\mathbb{Z})$, a and $d = \pm 1$ and so $\Im g(z) = \Im z$. This is impossible because z is not in Ω, but $g(z) \in \Omega$. Since g is only determined up to ± 1, we can, if necessary, multiply by -1 to make $c > 0$.

Now $|cz+d|^2 = (c\alpha+d)^2 + (cy)^2$ and, since the arithmetic mean is greater than or equal to the geometric mean, we see that

$$(c\alpha+d)^2 + (cy)^2 \geq 2|c\alpha+d|cy\,.$$

So $\frac{y}{2|c\alpha+d|cy} \geq \frac{\sqrt{3}}{2}$. Cancelling the y's and dividing by $c > 0$ gives

$$|\alpha - (-d/c)| \leq \frac{1}{\sqrt{3}c^2}\,.$$

5.3 Annuli

Now we use the principle of the argument together with certain other facts to study conformal mappings of annuli.

Let $A(r,\mathcal{R}) = \{z \in \mathbb{C} : r < |z| < \mathcal{R}\}$, where $r > 0$. This is a domain, but of course it is not simply connected. Its fundamental group is $\mathbb{Z}$.

Theorem 5.3.1 *$A(r_1,\mathcal{R}_1)$ and $A(r_2,\mathcal{R}_2)$ are holomorphically equivalent if and only if $\frac{\mathcal{R}_1}{r_1} = \frac{\mathcal{R}_2}{r_2}$.*

Proof. If we normalize things by taking $r_1 = 1 = r_2$, then $A(1,\mathcal{R}_1)$ and $A(1,\mathcal{R}_2)$ are holomorphically equivalent if and only if $\mathcal{R}_1 = \mathcal{R}_2$. This means there is a continuum of non-holomorphically equivalent domains here. This is in contrast to the simply connected case (see Theorem 5.4.1) where all domains $\neq \mathbb{C}$ are equivalent.

Let $\lambda > 0$ and consider the entire function $z \mapsto \lambda z$. This is a holomorphic function on $A(r_1,\mathcal{R}_1)$ which maps $A(r_1,\mathcal{R}_1)$ onto $A(\lambda r_1,\lambda\mathcal{R}_1)$. Taking $\lambda = \frac{1}{r_1}$ we see $A(r_1,\mathcal{R}_1)$ is equivalent to $A(1,\frac{\mathcal{R}_1}{r_1})$ and similarly $A(r_2,\mathcal{R}_2)$ is equivalent to $A(1,\frac{\mathcal{R}_2}{r_2})$. Hence, if $\frac{\mathcal{R}_1}{r_1} = \frac{\mathcal{R}_2}{r_2}$, then $A(r_1,\mathcal{R}_1)$ and $A(r_2,\mathcal{R}_2)$ are holomorphically equivalent. To prove our theorem in the other direction, we may assume $r_1 = 1 = r_2$, and show that if $A(1,\mathcal{R}_1) = A_1$ and $A(1,\mathcal{R}_2) = A_2$ are holomorphically equivalent, then $\mathcal{R}_1 = \mathcal{R}_2$ (cf. Figure 5.2).

We first need to know something about boundary behavior, namely, Theorem 2.1, p. 20 of [8]: Let $f : D \to \Omega$ be a conformal map, where D is the unit disk, Ω is a simply connected bounded domain with $\partial\Omega$ *locally*

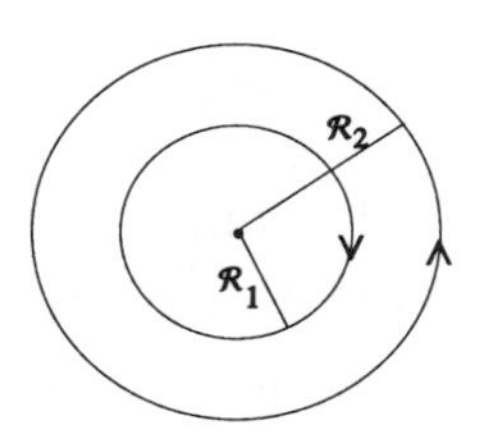

Figure 5.2

connected, then f extends continuously to ∂D. In particular, taking a straight line in A_1 joining the inner and outer circles and its image under f in A_2, it follows that f extends to a homeomorphism of the respective boundaries. (However, we could go with the present argument as it stands, simply by considering maps which are both holomorphic equivalences and homeomorphisms of the respective boundaries).

Let $f : \Omega_1 \to \Omega_2$ be a conformal equivalence between bounded domains in $\mathbb{C} = \mathbb{R}^2$ (actually this holds in $\mathbb{R}^n$). Then f extends to a homeomorphism $\phi : \partial(\Omega_1) \to \partial(\Omega_2)$ of the boundaries.

In particular, here ϕ maps each boundary component of A_1 onto some boundary component of A_2. Thus, if z_j is a sequence of points in A_1, where $|z_j| \to 1$, then $|f(z_j)| \to 1$ or to $\mathcal{R}_2$. In the latter case, we simply replace f by $\frac{\mathcal{R}_2}{f}$. Since $|f(z)| > 1$, $f \neq 0$, so this is permissible. We leave it to the reader to check that because of the properties of f, this new function is a holomorphic equivalence between A_1 and A_2 which maps the inner (respectively outer) boundary to the inner (respectively outer) boundary. Thus we may assume that as $|z_j| \to 1$, $|f(z_j)| \to 1$ and, as $|z_j| \to \mathcal{R}_1$, $|f(z_j)| \to \mathcal{R}_2$.

The key here is to let α be the positive real number $\frac{\log \mathcal{R}_2}{\log \mathcal{R}_1}$ and

$$\omega(z) = \log|f(z)| - \alpha \log|z|,$$

for $z \in A_1$. By Corollary 2.5.5, ω is a harmonic function on A_1. Since f extends to the boundary, so does ω via the same definition as above.

Therefore, if $|z| = 1$, then $\omega(z) = \log 1 - \alpha \log 1 = 0$ and also if $|z| = \mathcal{R}_1$, then $\omega(z) = \log \mathcal{R}_2 - \alpha \log \mathcal{R}_1 = 0$. Since ω is $\equiv 0$ on the boundary of A_1, it follows from Corollary 3.2.4 that ω is $\equiv 0$ on A_1

itself. This means $\log|f(z)| = \alpha \log|z|$ for all $z \in A_1$. Alternatively, $|f(z)| = e^{\log|z|\alpha} = |z|^\alpha$, for all $z \in A_1$.

Now choose a disk $D \subseteq A_1$. Since $f \neq 0$ and holomorphic on D and the latter is simply connected, $f(z) = e^{g(z)}$ for some holomorphic function g on D. Now, by Corollary 1.3.2, $|\frac{1}{z}e^{\frac{g(z)}{\alpha}}| = |\frac{1}{z}| \cdot |e^{\frac{g(z)}{\alpha}}|$. But the latter is

$$\frac{1}{|z|} \cdot |(e^{g(z)})^{\frac{1}{\alpha}}| = \frac{1}{|z|} \cdot |f(z)^{\frac{1}{\alpha}}| = \frac{1}{|z|}|z| = 1 .$$

Thus for all $z \in D$, we know $e^{\frac{g(z)}{\alpha}} = \lambda(z)z$, where $|\lambda(z)| = 1$. Now the point is that actually $\lambda(z)$ is constant, i.e. is independent of $z \in D$. Since $\lambda(z)$ has constant modulus, we need only see that it has constant argument.

First notice that $e^{\frac{g(z)}{\alpha}}$ has non-zero derivative everywhere on D. This is because $\frac{d}{dz}e^{\frac{g(z)}{\alpha}} = e^{\frac{g(z)}{\alpha}}\frac{g'(z)}{\alpha}$. Therefore, for this it is sufficient to see that $g'(z) \neq 0$ on D. But since $f(z) = e^{g(z)}$, we get $f'(z) = f(z)g'(z)$. Now $f \neq 0$ on all of A_1 and since f is an invertible holomorphic map also $f' \neq 0$ on all of A_1. Therefore, these hold on D and it follows that $g' \neq 0$ on D.

Since $e^{\frac{g(z)}{\alpha}}$ is regular and holomorphic on D, it is also conformal on D by Theorem 1.6.3. Let a be the center of D. It follows that

$$\arg e^{\frac{g(z)}{\alpha}} - \arg e^{\frac{g(a)}{\alpha}} = \arg z - \arg a .$$

That is,

$$\arg e^{\frac{g(z)}{\alpha}} - \arg z = \arg e^{\frac{g(a)}{\alpha}} - \arg a ,$$

a constant. Thus we get $e^{\frac{g(z)}{\alpha}} = \lambda z$, $z \in D$ where $|\lambda| = 1$. Differentiating this equation on D gives $e^{\frac{g(z)}{\alpha}}(\frac{g'(z)}{\alpha}) = \lambda$. After cancelling the λ, we get $g'(z) = \frac{\alpha}{z}$, for $z \in D$. Going back to f we have $\frac{f'(z)}{f(z)} = \frac{\alpha}{z}$, for all $z \in D$. But this involves only f and α. These quantities are given and are defined on all of A_1. Since the latter equation holds for all subdisks of A_1 and these fill out A_1 it follows that actually

$$\frac{f'(z)}{f(z)} = \frac{\alpha}{z} ,$$

for all $z \in A_1$. Now choose any circle γ_* centered at 0 with radius between 1 and $\mathcal{R}_1$. Integrating this last equation over γ_* we get

$$\frac{1}{2\pi i}\int_{\gamma_*}\frac{f'(z)}{f(z)}dz = \frac{1}{2\pi i}\int_{\gamma_*}\frac{\alpha}{z}dz\,.$$

As we know, the right hand side is α. The left hand side is given by the formula for variation of the argument (cf. Theorem 4.2.7) which is independent of whether the region (or the interior of γ_*) is simply connected, or not. Thus we see $\frac{1}{2\pi}\Delta_{\gamma_*}(f) = \alpha$. On the other hand, since f is 1:1 the image of γ_* under f is a simple closed curve and therefore has index ± 1, depending on whether f preserves or reverses orientation. So $\alpha = \pm 1$, but since α is positive, $\alpha = 1$. Hence $\log \mathcal{R}_2 = \log \mathcal{R}_1$ and $\mathcal{R}_2 = \mathcal{R}_1$. □

Corollary 5.3.2 *If $A = A(r, \mathcal{R})$, then* Aut(A) *is just the rotation group.*

Proof. Clearly the group of rotations acts on A by holomorphic automorphisms. Conversely, let f be such an automorphism. By the argument above we can assume $r = 1$ and since here $\mathcal{R}_1 = \mathcal{R}_2$, we know $\alpha = 1$. Since in general on a disk D we have $e^{\frac{g(z)}{\alpha}} = \lambda z$, here we get $e^{g(z)} = \lambda z$. That is, $f(z) = \lambda z$, for $z \in D$, where $|\lambda| = 1$. Now $\lambda = \frac{f(z)}{z}$ and so is independent of D. Arguing as before, we see that $f(z) = \lambda z$, for all $z \in A$. □

5.4 The Riemann mapping theorem for planar domains

The purpose of this section is to prove the Riemann mapping theorem for planar domains. Here we denote by D the interior of the unit disk.

Theorem 5.4.1 *Let $\Omega \neq \mathbb{C}$ be a simply connected domain and $a \in \Omega$ be fixed. Then there exists a unique holomorphic mapping $f : \Omega \to D$, which is bijective (and biholomorphic) such that $f(a) = 0$ and $f'(a) > 0$.*

Before turning to the proof we remark that, in particular, this tells us that any two simply connected domains, neither of which is $\mathbb{C}$, must be holomorphically equivalent since each is equivalent to D. For example, as we saw earlier, the upper half plane is equivalent to D via the Cayley transform. The requirement that $\Omega \neq \mathbb{C}$ is necessary since there can be no non-trivial holomorphic map $g = f^{-1} : \mathbb{C} \to D$. Such a map would be a bounded entire function. By Liouville's theorem it would have to be constant and therefore its derivative identically zero. However, g is invertible and therefore its derivative is everywhere $\neq 0$. (Notice, however, that these domains are real C^{∞} diffeomorphic.) Earlier, we also saw that if a domain is not simply connected, such as in the case of an annulus, the situation is completely different. Two annuli are holomorphically equivalent if and only if they have the same ratio of inner and outer radii.

Proof. First we prove uniqueness. Suppose f and g are two such maps. Then $f \circ g^{-1}$ is a holomorphic automorphism of D. Moreover, $f \circ g^{-1}(0) = f(a) = 0$. It follows from our study of the group $\mathrm{Aut}(D)$ that $f \circ g^{-1}$ is a rotation, i.e. $f \circ g^{-1}(z) = \lambda z$, for $z \in D$. Differentiating, we get $f'(g^{-1}(z))\frac{1}{g'(g^{-1}(z))} = \lambda$. Evaluating at $z = 0$ yields $\frac{f'(a)}{g'(a)} = \lambda$. But since $f'(a)$ and $g'(a) > 0$, so is their quotient. On the other hand, $|\lambda| = 1$. This means $\lambda = 1$ and so $f = g$.

We now turn to existence. Let $\mathcal{F}$ denote the set of all holomorphic functions $f : \Omega \to D$, which are injective with $f(a) = 0$ and $f'(a) > 0$.

As a first step we show (as would have to be the case if the theorem were true) that $\mathcal{F}$ is not empty. Since $\Omega \neq \mathbb{C}$, choose $b \in \mathbb{C} - \Omega$ and consider the function $z - b$ on Ω. As this function is holomorphic and never zero on Ω and our domain is simply connected, we know $z - b = e^{g(z)}$, where g is holomorphic on Ω. Let $f(z) = e^{\frac{g(z)}{2}}$; then f is also holomorphic and, moreover, $f^2(z) = z - b$ on Ω. Suppose $f(z_1) = f(z_2)$, where z_1 and $z_2 \in \Omega$. Squaring we get $z_1 - b = z_2 - b$ and so $z_1 = z_2$. Thus f is 1:1 on Ω. In particular, $f' \neq 0$ everywhere on Ω and so f is an open map. Hence $f(\Omega)$ is a non-trivial open set in $\mathbb{C}$. In particular, $f(\Omega)$ contains a disk $D(f(a), r)$, where $r > 0$. Now $f(\Omega)$ and $D(-f(a), r)$ are disjoint. If not, there is some $z \in \Omega$ where

$|f(z) + f(a)| < r$. That is, the distance from $-f(z)$ to $f(a)$ is less than r. But then $-f(z) \in D(f(a), r) \subseteq f(\Omega)$. Thus $-f(z) = f(z_1)$ for some $z_1 \in \Omega$. Squaring and arguing as above tells us $z = z_1$. Hence $-f(z) = f(z)$ so $f(z) = 0$ for this z. On the other hand, $f^2(z) = z - b$ for all $z \in \Omega$. It follows that $b = z$, so $b \in \Omega$. This contradiction proves $f(\Omega)$ and $D(-f(a), r)$ are disjoint.

We now consider stereographic projection. Let τ be a Möbius transformation that takes $S^2 - D(-f(a), r)$ onto D. Here is how to do this: See exercises at end of this section for definitions.

1. Translate $D(-f(a), r)$ so as to have its center at the origin.

2. Stretch or shrink it so that under stereographic projection (or rather its inverse) the disk maps exactly to the Southern hemisphere.

3. Then perform inversion in the sphere interchanging the Northern and Southern hemispheres and leaving the equator fixed. (In other words, perform inversion in the circle corresponding to the Southern hemisphere.) This gives a map between $D(-f(a), r)$ and the Northern hemisphere taking the boundary onto the equator.

4. Now apply stereographic projection to $S^2 - \overline{D(-f(a), r)}$, i.e. to the Southern hemisphere, and get a disk centered at the origin.

5. Finally, shrink or stretch this to get the unit disk.

Now let $f_1 = \tau \circ f$. Then f_1 is also holomorphic and 1:1. However, since $f(\Omega)$ and $D(-f(a), r)$ are disjoint, $f(\Omega) \subseteq S^2 - D(-f(a), r)$, and so, $f_1(\Omega) = \tau \circ f(\Omega) \subseteq D$. Let $\alpha = f_1(a)$ and $f_2(z) = \frac{f_1(z) - \alpha}{1 - \bar{\alpha} f_1(z)}$. This map is the composition of f_1 with an automorphism of the disk of the form translation by α . Notice that since f_1 takes values in D, we know $|\alpha| < 1$. Hence f_2 is also holomorphic, 1:1, and takes values in the unit disk but, in addition, $f_2(a) = 0$. Finally, let $f_3(z) = cf_2(z)$, where $|c| = 1$, c to be determined later. Since $|c| = 1$ and f_2 takes values in the unit disk so does f_3. In fact, f_3 shares all the other properties of f_2. Since $f_3'(z) = cf_2'(z)$, evaluating at $z = a$ and taking into account that

$f_2'(a) \neq 0$, choose c so that $cf_2'(a) > 0$, i.e. rotate $f_2'(a)$ so it matches up with the real axis. Evidently, $f_3 \in \mathcal{F}$.

Next we work in the space of complex valued functions on Ω, where the topology is that of uniform convergence on compacta and we strengthen the statement that $\mathcal{F}$ is not empty by showing $\bar{\mathcal{F}} = \mathcal{F} \cup \{0\}$.

To see this let $f_n \to f$ uniformly on compacta, where $f_n \in \mathcal{F}$. Since each f_n is holomorphic so is f. Moreover, since $f_n \to f$ pointwise and each f_n takes values in D, it follows that f takes values in $\bar{D}$. Also, because $f_n(a) \to f(a)$ and all $f_n(a) = 0$ it follows that $f(a) = 0$. Further, $f_n' \to f'$ uniformly on compacta so, in particular, $f_n'(a) \to f'(a)$. Since each $f_n'(a) > 0$, it also follows that $f'(a) \geq 0$.

Then we show either f is 1:1 on Ω, or f is $\equiv 0$ on Ω. Let z_1 and $z_2 \in \Omega$ be distinct points, with $f(z_2) = f(z_1)$. We will show that if f is not $\equiv 0$, then $z_2 = z_1$. Since these are arbitrary points in Ω this would prove f is 1:1. For every positive integer n, let $w_n = f_n(z_1)$. Choose a compact ball $C = C(z_2, r)$ in Ω centered at z_2 and so that z_1 is not in C. For each n, $f_n(z) - w_n$ does not vanish on C. If it did, then for that n, $f_n(z) = f_n(z_1)$, where $z \in C$. Since each f_n is 1:1 we would have $z = z_1$, which is a contradiction, since z_1 is not in C. On the other hand, $f_n(z_1) = w_n$ converges to $w = f(z_1)$. Hence $f_n(z) - f_n(z_1) = f_n(z) - w_n$ which tends to $f(z) - w$ on the compact set $C \cup \{z_1\}$. Since for each n, $f_n(z) - w_n$ does not vanish on C and we can choose $r > 0$ as small as we like, it follows from Hürwitz' theorem that either $f(z) - w$ never vanishes on C, or $f(z) - w$ is $\equiv 0$ on a sequence of points tending to z_2 and, therefore, by the identity theorem is $\equiv 0$ on C itself. In the latter case, f is constant on a ball and, therefore, again by the identity theorem, is constant on all of Ω. Because $f(a) = 0$, f is $\equiv 0$. In the other alternative, since $f(z) - w$, i.e. $f(z) - f(z_1)$ never vanishes on C. In particular, $f(z_2)$ cannot equal $f(z_1)$, because then this function would definitely vanish at $z_2 \in C$. This contradiction proves f is 1:1.

Thus f is either 1:1 or $\equiv 0$. In the former case, $f' \neq 0$ at every point and so f is an open map. Hence, $f(\Omega)$ is an open subset of $\bar{D}$ and, therefore, is contained in its interior, namely D. Also in this case $f'(a) > 0$ and so $f \in \mathcal{F}$. Hence $\bar{\mathcal{F}} \subseteq \mathcal{F} \cup \{0\}$. To prove the converse we just need to show that $0 \in \bar{\mathcal{F}}$. Let $f \in \mathcal{F}$ ($\mathcal{F}$ is non-empty) and let $\delta_n = \frac{1}{n} f$. Then for any compact set, K, in Ω $\| \frac{1}{n} f \|_K = \frac{1}{n} \| f \|_K$, which clearly tends to zero. Thus δ_n is a sequence tending to zero. But these δ_n lie in $\mathcal{F}$ (since $f \in \mathcal{F}$ so is cf for any c where $0 < c \leq 1$). Thus $\bar{\mathcal{F}} = \mathcal{F} \cup \{0\}$.

Finally, we show that for a well chosen $f \in \mathcal{F}$, $f(\Omega) = D$. This would complete the proof since f^{-1} is automatically holomorphic. Consider the functional on $H(\Omega)$, the holomorphic functions on Ω, given by $f \mapsto f'(a)$. This is a continuous functional since if $f_n \to f$ uniformly on compacta, then $f'_n \to f'$ uniformly on compacta and, in particular, $f'_n(a) \to f'(a)$. Now we will see that $\bar{\mathcal{F}}$ is compact because by Montel's theorem we need only check that $\mathcal{F}$ is locally bounded. Indeed, $\mathcal{F}$ is globally bounded since $\| f \|_\Omega \leq 1$, because $f(\Omega) \subseteq D$. By compactness, choose an $f \in \bar{\mathcal{F}}$ maximizing $f'(a)$. In particular, $f'(a) \geq g'(a)$ for all $g \in \mathcal{F}$.

We show $f(\Omega) = D$. If not, then there is some $w \in D - f(\Omega)$. Now $w \neq 0$ since $f(a) = 0$. Consider the function given by

$$T_w(z) = \frac{f(z) - w}{1 - \bar{w} \cdot f(z)}, z \in \Omega.$$

This is the composition of f with a holomorphic automorphism of D and so continues to take values in D. Notice that $1 - \bar{w} \cdot f(z)$ never vanishes on Ω. For if it did, then $f(z) = \frac{1}{\bar{w}}$ and since $w \in D$ this would force $f(z)$ to be outside of D, a contradiction, since $f(\Omega) \subseteq D$. Notice also that $T_w(z)$ never vanishes since $w \neq f(z)$ for any $z \in \Omega$. Now because $T_w(z)$ is a holomorphic function on a simply connected domain it has a holomorphic square root, $h(z)$.

$$h^2(z) = \frac{f(z) - w}{1 - \bar{w} \cdot f(z)}, z \in \Omega,$$

and since $h^2(z) \in D$ we know $|h(z)| < 1$. Differentiating and an easy calculation shows that

$$2h(z)h'(z) = \frac{(1 - |w|^2) f'(z)}{(1 - \bar{w} \cdot f(z))^2}.$$

Evaluating this at $z = a$ and taking into account $f(a) = 0$ gives $2h(a)h'(a) = (1 - |w|^2) f'(a)$. Since $|w|^2 < 1$ and $f'(a) > 0$ it follows that $h(a)h'(a) \neq 0$ and hence that both $h(a)$ and $h'(a) \neq 0$. Moreover, $1 - \overline{h(a)} \cdot h(z)$ is never zero on Ω since both $|h(a)| < 1$ and $|h(z)| < 1$. Let

$$g(z) = \frac{|h'(a)|}{h'(a)} \frac{h(z) - h(a)}{1 - \overline{h(a)} \cdot h(z)}, z \in \Omega.$$

Then g is holomorphic on Ω and $g(a) = 0$. Since $|\frac{|h'(a)|}{h'(a)}| = 1$ and because $T_{h(a)}$ is an automorphism of D we get $|\frac{h(z)-h(a)}{1-\overline{h(a)} \cdot h(z)}| < 1$. Hence $|g(z)| < 1$.

We now show g is 1:1. Suppose $g(z) = g(z')$, where z and $z' \in \Omega$.

$$\frac{|h'(a)|}{h'(a)} \frac{h(z) - h(a)}{1 - \overline{h(a)} \cdot h(z)} = \frac{|h'(a)|}{h'(a)} \frac{h(z') - h(a)}{1 - \overline{h(a)} \cdot h(z')}.$$

Hence $\frac{h(z)-h(a)}{1-\overline{h(a)} \cdot h(z)} = \frac{h(z')-h(a)}{1-\overline{h(a)} \cdot h(z')}$. That is to say, $T_{h(a)}(h(z)) = T_{h(a)}(h(z'))$. Since $T_{h(a)}$ is 1:1, $h(z) = h(z')$ and, therefore, $h^2(z) = h^2(z')$. This means $\frac{f(z)-w}{1-\overline{w} \cdot f(z)} = \frac{f(z')-w}{1-\overline{w} \cdot f(z')}$. Thus $T_w(f(z)) = T_w(f(z'))$. Since T_w is 1:1, $f(z) = f(z')$. But f is itself 1:1, so $z = z'$.

A direct calculation shows that

$$g'(z) = \frac{|h'(a)|}{h'(a)} \frac{h'(z)(1 - |h(a)|^2)}{(1 - \overline{h(a)} \cdot h(z))^2}.$$

Hence,

$$g'(a) = \frac{|h'(a)|}{(1 - |h(a)|^2)}.$$

Since $h^2(a) = -w$, $|h(a)|^2 = |w|$ and so $g'(a) = \frac{|h'(a)|}{1-|w|}$. But since $2h(a)h'(a) = (1 - |w|^2)f'(a)$, taking absolute values and keeping in mind that $g'(a)$ and $f'(a)$ are both positive gives us

$$g'(a) = \frac{(1 - |w|^2)f'(a)}{(1 - |w|)2\sqrt{|w|}} = \frac{1 + |w|}{2\sqrt{|w|}} f'(a).$$

Letting $t = \sqrt{|w|}$, we see $g'(a) = \frac{1+t^2}{2t} f'(a)$, where $0 < t < 1$. Since for such t, $\frac{1+t^2}{2t} > 1$, $g'(a) > f'(a)$. This both shows that $g'(a) > 0$ so $g \in \mathcal{F}$ and gives a contradiction since $f'(a)$ is the largest such. $\square$

We remark that if Ω is a simply connected bounded domain with $\partial\Omega$ locally connected, then Theorem 2.1, p.20 of [8] mentioned in the previous section tells us that f actually extends to a homeomorphism of the respective boundaries.

Exercise 5.5 1. *Show that if t is real and $t \neq 1$, then $\frac{1+t^2}{2t} > 1$.*

2. *Show if x and y are real and non negative, then $\frac{x+y}{2} \geq \sqrt{xy}$, with equality only if $x = y$.*

3. *Prove the Möbius group is generated by translations ($z \mapsto z + b$, where $b \in \mathbb{C}$), dilations ($z \mapsto az$, where $a \in \mathbb{C}$, $a \neq 0$) and inversion ($z \mapsto \frac{1}{z}$).*

4. *Given four distinct points $z_1, z_2, z_3, z_4 \in \mathbb{C}$ we define their* cross ratio *by*

$$(z_1, z_2, z_3, z_4) = \frac{(z_1 - z_4)(z_3 - z_2)}{(z_1 - z_2)(z_3 - z_4)}.$$

Prove by direct calculation that the cross ratio is an invariant of the Möbius group G. That is, $(g(z_1), g(z_2), g(z_3), g(z_4)) = (z_1, z_2, z_3, z_4)$ for every $g \in G$.

5. *Given two sets of three distinct points, z_1, z_2, z_3 and $w_1, w_2, w_3 \in \mathbb{C}$ show there is a unique fractional linear transformation $g \in G$ such that $g(z_j) = w_j$ for all $j = 1, 2, 3$. Suggestion: If $z \neq z_j$ for $j = 1, 2, 3$ and $g \in G$, then we must have $(w_1, w_2, w_3, g(z)) = (z_1, z_2, z_3, z)$. Now consider what happens as $z \to z_j$ for one of the j. Solve the equation above for $g(z)$ and note that (z_1, z_2, z_3, z) is a fractional linear transformation of z.*

6. *Show the Möbius group takes circles to circles.*

7. *Prove the following generalization of the Schwarz lemma. (The reader should check that this is a generalization). Let D be the unit disk and $f : D \to D$ be a holomorphic map, with $f(z_0) = w_0$. Then*

(a) *For all* $z \in D$, $|\frac{f(z)-w_0}{1-\overline{w_0}f(z)}| \leq |\frac{z-z_0}{1-\overline{z_0}z}|$. *If equality holds at any point other than* z_0, *then* f *is a fractional linear transformation.*

(b) $|f'(z_0)| \leq \frac{1-|f(z_0)|^2}{1-|z_0|^2}$. *If equality holds, then* f *is a fractional linear transformation.*

Chapter 6

Applications of Complex Analysis to Lie Theory

6.1 Applications of the identity theorem: Complete reducibility of representations according to Hermann Weyl and the functional equation for the exponential map of a real Lie group

We first prove that holomorphic finite dimensional representations of a Lie group whose Lie algebra has a compact real form are completely reducible. This uses the so-called unitarian trick and was actually the first proof of this result. Of course, there is now a more general algebraic proof of this fact, but the present one retains great appeal to the author. Here a representation of the group G on a finite dimensional complex vector space V is a group homomorphism $\rho : G \to \mathrm{GL}(V)$, where $\mathrm{GL}(V) = \mathrm{GL}(n, \mathbb{C})$ is the group of $n \times n$ invertible matrices with complex entries. We say ρ is holomorphic if each of these coordinate functions is holomorphic. For an exposition of this and any other facts concerning Lie groups see G. Hochschild [4].

The proof of Theorem 6.1.2 below requires the following simple lemma. Here, a complex valued function ϕ on a domain in $\mathbb{C}^n$ is called holomorphic if it is holomorphic in each variable separately, while holding all the others fixed.

Lemma 6.1.1 *Let $\phi : \mathbb{C}^n \to \mathbb{C}$ be an entire function which vanishes identically on $\mathbb{R}^n$. Then $\phi \equiv 0$.*

Proof. $\phi(z_1, \dots, z_n)$ vanishes when all z_i are real, so consider $\phi(z_1, x_2 \dots, x_n)$ where the x_i are real. This is an entire function of z_1 and vanishes on the real axis. By the identity theorem, it vanishes identically. Let z_1 be fixed, but arbitrary and consider $\phi(z_1, z_2, x_3 \dots, x_n)$, where the x_i are real. This is an entire function of z_2 which vanishes when z_2 is real and, therefore, identically in z_2. Continuing by induction, we see that $\phi(z_1, \dots, z_n) \equiv 0$. □

It also requires just two facts of Lie theory.

1. Let ρ be a representation of G on V and ρ° be its differential representation on the Lie algebra $\mathfrak{g}$. If H is a closed subgroup of G with Lie algebra $\mathfrak{h}$, then a subspace W of V is H-invariant if and only if it is $\mathfrak{h}$-invariant.

2. The Lie algebra $\mathfrak{g}$ of a complex semisimple Lie group G has a compact real form $\mathfrak{k}$ and the Lie subgroup K of G with Lie algebra $\mathfrak{k}$ is compact.

Theorem 6.1.2 *Let G be a complex connected Lie group whose Lie algebra has a compact real form, $\mathfrak{k}$. Then every finite dimensional holomorphic representation is completely reducible.*

Proof. Let ρ be a holomorphic representation of G on V. We will show that if W is a K-invariant subspace of V, then it is actually G-invariant. This would imply the following:

1. If ρ is irreducible, then so is $\rho|K$.

2. If $\rho|K$ is completely reducible, then so is ρ.

1. If not, then since K is compact, $\rho|K = \Sigma\rho_i$, a direct sum of irreducibles. Each of the corresponding subspaces V_i is K and, therefore, G-invariant. Hence ρ is reducible, a contradiction.

2. Let W be a G-invariant subspace of V. Then W is K-invariant. Since K is compact, there is a complementary K-invariant subspace W' which would be therefore G-invariant. Thus ρ would be completely reducible.

Therefore, it only remains to show that if W is a $\mathfrak{k}$-invariant $\mathbb{C}$ subspace of V, then it is $\mathfrak{g}$-invariant. Let $\lambda \in (V/W)^*$, the $\mathbb{C}$-dual of V/W, and $w \in W$. Then for $k \in \mathfrak{k}$ we know $\rho_k(w) \in W$ and, hence, $\lambda(\rho_k(w)) = 0$. For $X \in \mathfrak{g}$ let $\phi(X) = \lambda(\rho_X(w))$, where $w \in W$ and $\lambda \in (V/W)^*$. Then $\phi : \mathfrak{g} \to \mathbb{C}$ is an entire function which vanishes on $\mathfrak{k}$ and hence, by Lemma 6.1.1, it vanishes on all of $\mathfrak{g}$. Since this is true for all $w \in W$ and all $\lambda \in (V/W)^*$, and the dual space separates the points, it follows that W is $\mathfrak{g}$-invariant. □

We now apply the identity theorem to derive the functional equation for the exponential map of a connected real Lie group, G. As above, $\mathfrak{g}$ is the Lie algebra of G. Here $\exp : \mathfrak{g} \to G$ denotes the exponential map. A well known property of this map is the fact that if X and $Y \in \mathfrak{g}$ commute, i.e., if $[X, Y] = 0$ in $\mathfrak{g}$, then

$$\exp X \cdot \exp Y = \exp(X + Y)\,. \tag{6.1}$$

To see why this is so we recall the Baker-Campbell-Hausdorff (BCH) formula.

$$\exp(X) \cdot \exp(Y) = \exp\left(X + Y + \frac{1}{2}[X, Y] + \cdots\right), \tag{6.2}$$

where the right side is an absolutely convergent series, involving higher order brackets of X and Y, and (6.2) is valid for all *small* X and Y. That is, X and Y lie in a small ball about zero in $\mathfrak{g}$.

Now if X and Y commute, so do tX and tY for all real t. Take $|t|$ small enough so that the BCH formula applies to tX and tY. Hence, by Eq. (6.2), since all the other terms in the formula are zero,

$$\exp tX \cdot \exp tY = \exp t(X + Y)\,, \tag{6.3}$$

for all small t. Now the exponential function as well as multiplication in the group are real analytic and $\exp tX$ is defined for all t. Also, by the chain rule, the composition of real analytic functions is again real analytic. Hence, by the identity theorem for real analytic functions (see remarks at the end of Section 3.3), this holds for all t. Taking $t = 1$ gives the result.

6.2 Application of residues: The surjectivity of the exponential map for U(p,q)

Given a connected Lie group G with Lie algebra $\mathfrak{g}$ we shall say G is *exponential* if $\exp : \mathfrak{g} \to G$ is surjective, or equivalently if each point of G lies on a one parameter subgroup.

Here we consider $\mathrm{U}(p,q)$, the unitary group with respect to a non-degenerate, but indefinite form. Although there are some vagaries in their argument, the authors in [11], using complex analysis, prove that this non-compact Lie group is exponential. Our purpose here is to illustrate this application of complex analysis (Cauchy's integral formula and the residue theorem) to Lie group theory as well as to tighten up their proof. These results, as well as the exponentiality of a number of other classical groups, were also proved by D. Djokovic in [2] using different methods.

Let $n = p + q$ be positive integers and $\langle , \rangle_{p,q}$ be a Hermitian form of type p, q on $\mathbb{C}^n$. The group $\mathrm{U}(p,q)$ consists of all $\mathbb{C}$-linear maps on $\mathbb{C}^n$ which preserve this form. This is a connected, reductive, real linear Lie group, whose Lie algebra $\mathfrak{u}(p,q)$ consists of those matrices in $M_n(\mathbb{C})$ which are skew Hermitian with respect to $\langle , \rangle_{p,q}$.

Theorem 6.2.1 *The* $\mathrm{U}(p,q)$ *is exponential.*

Before proving this result we need a lemma. Henceforth we shall write $\langle , \rangle_{p,q}$ simply as $\langle , \rangle$.

Lemma 6.2.2 *For each* $g \in \mathrm{U}(p,q)$, *those eigenvalues of* g *not lying on the unit circle are symmetric with respect to it. That is, they come in pairs* λ, μ, *where* $\mu = \frac{\lambda}{|\lambda|^2}$.

Since for $\lambda \neq 0$, $\frac{\lambda}{|\lambda|^2} = \overline{(\lambda^{-1})}$, this function is precisely Schwarz reflection in the unit circle of $\mathbb{C}$.

Proof. Let $g \in \mathrm{U}(p,q)$ and λ be a fixed eigenvalue of g. Then $\det(g - \lambda I) = 0$. Multiplying by the invertible $g^{-1} = g^*$ we get $\det(I - \lambda g^*) = 0$. Since g is invertible, $\lambda \neq 0$, so multiplying this by λ^{-1} yields $\det(\lambda^{-1} I - g^*) = 0$. Hence $\det(g^* - \lambda^{-1} I) = 0$. Let T be any singular linear operator, then T^* is also singular. This is because with respect to the non-degenerate form, $\langle Tv, w\rangle = \langle v, T^*w\rangle$. If $Tv = 0$ for some non-zero v, then for all $w \in V$, $T^*(w)$ is orthogonal to that v, so $T^*(V)$ lies in a subspace of codimension at least one and, therefore, T^* is singular. Taking $T = g^* - \lambda^{-1} I$, we conclude $\det(g - \lambda^{-1})^* I = 0$. Since $\langle , \rangle$ is conjugate linear in the second variable, for any $\mu \in \mathbb{C}$, $(\mu I)^* = \bar{\mu} I$. Thus, $\det(g - \overline{\lambda^{-1}} I) = 0$ and hence $\overline{\lambda^{-1}}$ is also an eigenvalue of g. □

To prove this let $g \in \mathrm{U}(p,q)$ and $\lambda_1, \ldots \lambda_r$ be its distinct eigenvalues. By the lemma those eigenvalues not lying on the unit circle are symmetric with respect to it. We remark that we may replace g by wg where $w \in \mathbb{C}$ and $|w| = 1$. This is because $wg \in \mathrm{U}(p,q)$ and, if $wg = \exp X$, where X is skew Hermitian with respect to $\langle , \rangle$, then letting $z \in \mathbb{C}$ be pure imaginary with $e^z = w$ we see that $-zI + X$ is also skew Hermitian and $\exp(-zI + X) = \frac{1}{w} \exp X = g$. Choose a family γ_k of non-intersecting, positively oriented, small circles centered about each eigenvalue, so that for the pairs λ, μ off the unit circle the radii are chosen so that the function $z \mapsto (\bar{z})^{-1}$ takes γ_λ onto γ_μ (cf. Figure 6.1).

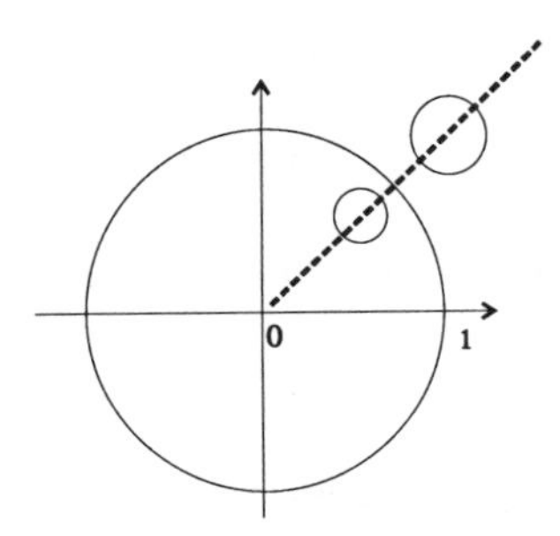

Figure 6.1

If necessary, by shrinking these circles further, subject to the above conditions, and rotating by a suitable w, as above, we can arrange that the negative real axis does not intersect any of these circles. Any point off this slit has a well defined argument between $-\pi$ and π. Choosing a branch, Log, of the log with respect to this slit, we now use complex functional calculus in this simply connected domain to define a linear operator, $\mathrm{Log}(g)$, by contour integration. We take any simple, closed, piecewise smooth curve γ in the domain containing all these circles in its interior and employ the Cauchy integral formula together with the residue theorem for the operator valued function Log.

$$\mathrm{Log}(g) = \frac{1}{2\pi i}\sum_{k=1}^{r}\int_{\gamma_k}(zI-g)^{-1}\,\mathrm{Log}(z)dz.$$

Then taking $X = \mathrm{Log}(g)$, evidently we have $\exp X = g$. We shall complete the proof by showing X is skew Hermitian, with respect to the form. To do so, we require the following lemma.

Lemma 6.2.3 *Let* $g \in \mathrm{GL}(n,\mathbb{C})$ *and* $z \in \mathbb{C}^\times$, *but off Spec* g. *Then*

$$z^{-2}(z^{-1}I - g^{-1})^{-1} = (g - zI)^{-1} + z^{-1}I\,.$$

Also, for z *off the slit,* $\mathrm{Log}(z^{-1}) = -\mathrm{Log}(z)$.

Proof. Evidently, $z^{-1}I - g^{-1} = z^{-1}(g - zI)g^{-1}$ so that since everything in sight commutes,

$$z^{-2}(z^{-1}I - g^{-1})^{-1} = z^{-2}z(g-zI)^{-1}g = z^{-1}g(g-zI)^{-1}\,.$$

But, since $z^{-1}g = I + z^{-1}(g - zI)$, we have

$$z^{-2}(z^{-1}I - g^{-1})^{-1} = (I + z^{-1}(g-zI))(g-zI)^{-1} = (g-zI)^{-1} + z^{-1}I\,.$$

Moreover, if $z = |z|e^{i\arg z}$, then $z^{-1} = |z|^{-1}e^{-i\arg z}$, so that

$$\mathrm{Log}(z) = \mathrm{Log}(|z|) + i\arg z$$

and $\mathrm{Log}(z^{-1}) = -\mathrm{Log}(|z|) - i\arg z = -\mathrm{Log}(z)$. □

Continuing the proof of the theorem, we know that

$$X^* = \mathrm{Log}(g)^* = -\frac{1}{2\pi i}\sum_{k=1}^{r}\int_{\gamma_k^*}(\bar{z}I - g^*)^{-1}\overline{\mathrm{Log}(z)dz}\,,$$

where γ_k^* means the conjugate of γ_k (with negative orientation). If we want to consider the circle, γ_k, to have negative orientation we shall write $\overline{\gamma_k}$. We now perform a change of variable under the orientation preserving map $w \mapsto \overline{w^{-1}}$. The circles, $\overline{\gamma_k}$, which do not intersect the unit circle, transform to the corresponding reflected circles, but also with acquire negative orientation, while the circles which meet the unit circle remain invariant, and also change orientation. Thus the total contour over which integration takes place remains invariant under this change of variables. Since $g^* = g^{-1}$, $\mathrm{Log}(\bar{z}) \mapsto \mathrm{Log}(z^{-1}) = -\mathrm{Log}(z)$ and $\overline{dz} = -z^{-2}dz$ we get

$$-\frac{1}{2\pi i}\sum_{j=1}^{r}\int_{\overline{\gamma_j}}(z^{-1}I - g^{-1})^{-1}(-\mathrm{Log}(z))(-z^{-2})dz\,.$$

Hence, after changing the orientation back to the positive one and cancelling out pairs of minus signs, X^* is given by

$$\frac{1}{2\pi i}\sum_{j=1}^{r}\int_{\gamma_j}(z^{-1}I - g^{-1})^{-1}\,\mathrm{Log}(z)z^{-2}dz\,.$$

Applying the lemma yields

$$\frac{1}{2\pi i}\sum_{j=1}^{r}\int_{\gamma_j}[(g - zI)^{-1} + z^{-1}I]\,\mathrm{Log}(z)dz\,.$$

Thus,

$$X^* = \frac{1}{2\pi i}\sum_{j=1}^{r}\int_{\gamma_j}(g - zI)^{-1}\,\mathrm{Log}(z)dz + \frac{1}{2\pi i}\sum_{j=1}^{r}\int_{\gamma_j}\frac{\mathrm{Log}(z)}{z}Idz\,.$$

Since the function $\frac{\mathrm{Log}(z)}{z}$ is holomorphic off the slit and the contour is closed, these last summands are zero. Writing $(g-zI)^{-1}$ as $-(zI-g)^{-1}$

and taking into account that the $\{\gamma_j\}$ are a permutation of the original $\{\gamma_k\}$ we get

$$X^* = -\frac{1}{2\pi i}\sum_{j=1}^{r}\int_{\gamma_j}(zI-g)^{-1}\,\mathrm{Log}(z)dz = -X\,.$$

Thus X is skew-Hermitian with respect to $\langle,\rangle_{p,q}$. This completes the proof. □

6.3 Application of Liouville's theorem and the maximum modulus theorem: The Zariski density of cofinite volume subgroups of complex Lie groups

The classical density theorem, due to A. Borel, deals with lattices in real semisimple groups without compact factors. An extension of this result to complex analytic groups below was first proved in [7]. Given a Lie group G and a closed subgroup H we say G/H has *finite volume* if there is a positive, finite, regular G-invariant measure on G/H, where G acts on the quotient space by left translation.

Theorem 6.3.1 *Let G be a connected complex Lie group and $\sigma : G \to \mathrm{GL}(V)$ a holomorphic representation of G on a finite dimensional complex vector space V. If H is a closed subgroup with G/H of finite volume, then each H-invariant subspace of V must also be G-invariant.*

Although we can not give the complete proof of Theorem 6.3.1 here, we can indicate the essentials, that is, how it depends on complex analysis. Here boundedness questions refer to any convenient Banach algebra norm $\|\cdot\|$ on $\mathrm{End}_{\mathbb{C}}(V)$. By a complex 1-parameter subgroup S of a complex analytic group G we mean a set of the form $S = \{\exp zX : z \in \mathbb{C}\}$, where $X \in \mathfrak{g}$, the Lie algebra of G. Notice that if ρ is a holomorphic representation of a complex analytic group G and W is an invariant subspace, then we also have a holomorphic representation of G on W. The proof of Theorem 6.3.1 rests on the following two facts about holomorphic representations. The holomorphic representations to which these observations are to be applied are $\rho = \wedge^r \sigma$, where $r = 1, \ldots, \dim V$.

1. The invariant subspace V_c of V consisting of vectors with bounded orbit is trivial.

2. For each one parameter subgroup S of G, if W is an S-invariant subspace of V, then either S acts on W by scalars, or else there is a sequence $\{g_k\}$ in $\rho(S)$ such that

$$\frac{\det(g_k|W)}{\| (g_k|W) \|^{\dim W}} \to 0.$$

Proof of the first statement. $W = V_c$ is a G-invariant subspace of V and on it G acts holomorphically as a bounded group. Therefore, it is sufficient to prove that a bounded holomorphic representation ρ of G on a complex vector space V is trivial. Since the 1-parameter subgroups generate G, to do this we may assume we are on a 1-parameter subgroup, S. Now the composition of holomorphic functions is holomorphic (Corollary 3.3.9). Hence $z \mapsto \rho(\exp zX) \subseteq \mathrm{GL}(V)$ is an entire matrix valued function. Since it is bounded, each matrix coefficient is a bounded entire function. By Liouville's theorem (Corollary 3.2.12) $\rho(\exp zX)$ must be constant. Taking $z = 0$, we see this constant is I and hence ρ is trivial on S.

Proof of the second statement. Let W be an $\exp zX$ invariant subspace of V. Then W is X-invariant and we may as well assume $W = V$. We show $\frac{\det g}{\|g\|^n} \to 0$ for some sequence of the g's in $\rho(S)$. Here $\rho(\exp zX) = \operatorname{Exp} z\rho^\circ(X)$, which we write henceforth as $\operatorname{Exp} zY$. But

$$\| \frac{g^{(n)}}{\det(g)} \| \leq \frac{\| g \|^n}{|\det(g)|},$$

so it suffices to show that $\| \frac{g^{(n)}}{\det(g)} \| \to \infty$. Now $g = \operatorname{Exp} zY$ so

$$\frac{g^{(n)}}{\det(g)} = \frac{\operatorname{Exp} nzY}{\det(\operatorname{Exp} zY)}.$$

This is an entire vector valued function of $z \in \mathbb{C}$. By the maximum principle (Corollary 3.7.4), it tends to ∞ as $|z|$ does, or it is constant.

In the latter case, $\frac{g^{(n)}}{\det(g)} = A \in \mathrm{End}_{\mathbb{C}}(V)$. Taking $g = 1$, we see that $A = I$ and $g^{(n)} = \det(g)I$. Since each (fixed) nth power of every element on the one parameter group acts as a scalar and every such element has an nth root, the whole one parameter group S acts as scalars.

Applying the first statement to the adjoint representation yields the following

Corollary 6.3.2 *A compact, connected, complex Lie group is abelian. In fact, it is an even dimensional torus.*

6.4 Applications of the identity theorem to differential topology and Lie groups

Theorem 6.4.1 *Let N be a connected real or complex analytic manifold and $f : N \to k$ be a real or complex analytic map, where $k = \mathbb{R}$ or $\mathbb{C}$, respectively. If f is non-constant, then $N(f) = \{p \in N : f(p) \neq 0\}$ is open and dense in N.*

Clearly $N(f)$ is open by continuity of the map. Because of the remarks made above about the relationship of real analytic functions to complex analytic functions, and because this holds in several variables just as well as one variable, the situation is more or less the same in the real and complex cases. Therefore, we shall deal only with the real case. We shall first prove this fact when N is Euclidean space, $\mathbb{R}^n$ and then deal with the general case.

Proof. If $N(f)$ were not dense there would be a non-trivial open set $U \subseteq \mathbb{R}^n$ on which $f \equiv 0$. Since U contains an open box, i.e. a product of n small open intervals (on which f also vanishes), we can apply a slightly modified form of Lemma 6.1.1 above to conclude $f \equiv 0$ on all of $\mathbb{R}^n$. This gives a contradiction.

Now let N be a connected analytic manifold. At each point $p \in N$ choose a neighborhood U_p about p which is analytically diffeomorphic to Euclidean space. Then $f|U_p$ is analytic on U_p and is non-constant for each $p \in N$. For otherwise for some p_0, $f|U_{p_0}$ would be constant. As in Chapter 2, let $p \in N$ and by connectedness join p to p_0 by a smooth curve lying in N. By compactness choose a finite number of points p_i on its trajectory such that each successive pair of the U_{p_i} have a non void intersection. By the identity theorem, f has the same constant value on $\cup U_{p_i}$ and, in particular, $f(p) = f(p_0)$. Thus f is constant, a contradiction. Therefore for each $p \in N$, $f|U_p$ is a non-constant analytic function. After identifying U_p with Euclidean space, from the Euclidean case above we conclude for each $p \in N$ that $N(f) \cap U_p$ is dense in U_p. Since the $\{U_p\}$ cover N it follows that $N(f)$ is dense in N. □

Now suppose $F : \mathbb{R}^n \to \mathbb{R}^n$ is a real analytic map and F_i are the component functions. In these coordinates the derivative of F at p is the $n \times n$ matrix

$$J_F(p) = \left(\frac{\partial F_i}{\partial x_j}(p)\right).$$

Corollary 6.4.2 *Let N be a connected real analytic manifold and $F : N \to N'$ a real analytic map, where N' is a real analytic manifold of the same dimension as N. If $\det J_F(p)$ is non-zero at some point p_0 of N, then the regular values of F, namely, $\{p : \det J_F(p) \neq 0\}$ is open and dense in N. The same holds for complex analytic maps on complex connected manifolds.*

Since det is a polynomial in the i, j coordinates with integer coefficients and therefore is analytic, and the composition of analytic functions is analytic, it follows that $f(p) = \det J_F(p)$ is an analytic function. By assumption f is non-zero at p_0. If it is zero anywhere else, it is non-constant and hence the corollary follows from Theorem 6.4.1. Otherwise, it must be non-zero everywhere and hence the regular values of f coincide with the whole space, N.

Corollary 6.4.3 *Let $\exp : \mathfrak{g} \to G$ be the exponential map of a real, or complex connected Lie group G. Then the set of regular values of $\exp$, i.e. the places in $\mathfrak{g}$ where $J(\exp)$ is non-singular, is open and dense.*

Because $\mathfrak{g}$ and G are real (respectively, complex) manifolds of the same dimension with $\mathfrak{g}$ connected, this follows immediately from

Corollary 6.4.2, since exp is analytic and a local diffeomorphism of a neighborhood about 0 in $\mathfrak{g}$ and hence $\det(J(\exp))$ is non-zero in some neighborhood of zero.

Corollary 6.4.4 *Let $k = \mathbb{R}$ or $\mathbb{C}$ and $M_n(k)$ be the $n \times n$ matrices over k. Then the places where $\det M_n(k) \to k$ is non-zero is open and dense in $M_n(k)$.*

Since det takes all values, this follows immediately from Corollary 6.4.2 by taking F to be the identity map.

Bibliography

[1] John B. Conway. *Functions of one complex variable.* Springer-Verlag, New York, second edition, 1978.

[2] Dragomir Ž. Djoković. On the exponential map in classical Lie groups. *J. Algebra*, 64(1):76–88, 1980.

[3] I. M. Gel′fand, M. I. Graev, and I. I. Pyatetskii-Shapiro. *Representation theory and automorphic functions.* W. B. Saunders Co., Philadelphia, Pa., 1969.

[4] G. Hochschild. *The structure of Lie groups.* Holden-Day Inc., San Francisco, 1965.

[5] Konrad Knopp. *Theory of Functions.* Part I, Dover, New York, 1952.

[6] Edwin E. Moise. *Geometric topology in dimensions* 2 *and* 3. Springer-Verlag, New York, 1977. Graduate Texts in Mathematics, Vol. 47.

[7] Martin Moskowitz. On the density theorems of Borel and Furstenberg. *Ark. Mat.*, 16(1):11–27, 1978.

[8] Ch. Pommerenke. *Boundary behaviour of conformal maps.* Springer-Verlag, Berlin, 1992.

[9] H. L. Royden. *Real analysis.* Macmillan Publishing Company, New York, third edition, 1988.

[10] Walter Rudin. *Real and complex analysis.* McGraw-Hill Book Co., New York, third edition, 1987.

[11] V. A. Yakubovich and V. M. Starzhinskii. *Linear differential equations with periodic coefficients.* 1, 2. Halsted Press [John Wiley & Sons], New York-Toronto, Ont., 1975. Translated from Russian by D. Louvish.

Index